T0218617

STUDYING GEOGRAPHY AT UNIVERSITY

Written by leading academics, this book is an invaluable 'how to …' guide to studying for a Geography degree. Written in a practical and conversational style, it offers important insights into how to succeed in the first year of your degree course, covering everything from how to succeed in assessments to how to decide where to live. Some of the information the book provides is academic and some of it is non-academic, as negotiating both is important in order to be successful in the first year of a Geography degree.

Studying Geography at University is ideal for those in the early stages of applying to university. Each chapter offers hints and tips and gives practical real-world insights into becoming a successful geography student that will enrich applications, open days and visit days. It is also possible to dip into the chapter summaries, 'What Do Students Say?' and 'Top Tip' boxes only. Written by current students, from a range of institutions, these provide unique insights into the book's key points. Current students should also keep and refer to the book as an invaluable guide through the first few months of their degree.

This guide is a must-read for anyone starting their studies in Human Geography, Physical Geography, Environmental Science or any other related subject at university.

Simon Tate is Professor of Pedagogy in Higher Education at Newcastle University, UK. Before taking up his current appointment, Simon taught Geography in comprehensive schools in the North East of England.

Peter Hopkins is Professor in Social Geography at Newcastle University, UK, where he has taught for over ten years.

STUDYING GEOGRAPHY AT UNIVERSITY

How to Succeed in the First Year of Your New Degree

SIMON TATE AND PETER HOPKINS

Routledge
Taylor & Francis Group

LONDON AND NEW YORK

First published 2021
by Routledge
2 Park Square, Milton Park, Abingdon, Oxon OX14 4RN

and by Routledge
52 Vanderbilt Avenue, New York, NY 10017

Routledge is an imprint of the Taylor & Francis Group, an informa business

British Library Cataloguing-in-Publication Data
A catalogue record for this book is available from the British Library

Library of Congress Cataloging-in-Publication Data
Names: Tate, Simon, author. | Hopkins, Peter, author.
Title: Studying geography at university: how to succeed in the first year of your new degree/Simon Tate and Peter Hopkins.
Description: Abingdon, Oxon; New York, NY: Routledge, 2020. | Includes bibliographical references and index.
Identifiers: LCCN 2020018875 (print) | LCCN 2020018876 (ebook) |
Subjects: LCSH: Geography–Study and teaching (Higher) | College students–Social conditions.
Classification: LCC G73 .T32 2020 (print) | LCC G73 .T32 2020 (ebook) | DDC 910.71/1–dc23
LC record available at https://lccn.loc.gov/2020018875
LC ebook record available at https://lccn.loc.gov/2020018876

ISBN: 978-0-8153-6968-4 (hbk)
ISBN: 978-0-8153-6969-1 (pbk)
ISBN: 978-1-351-16676-8 (ebk)

Typeset in Adobe Garamond Pro and Avenir
by Cenveo® Publisher Services

CONTENTS

FIGURES

TABLES

ACKNOWLEDGEMENTS

This book is better because of the many people who took time to read and comment upon our ideas at various stages. First, many thanks must go to our colleagues at Routledge for taking the time to consider our book proposal and to the constructive reviewers for their insights into our initial ideas. We are also indebted to the students and colleagues, from several universities, who read and commented upon early drafts of this book. Their views have helped to ensure that its content will be relevant to you, wherever you choose to study Geography. In particular, we are indebted to our friend and Physical Geography colleague, Dr Simon Drew, who wrote Chapter 21. As we are both Human Geographers, our conversations with Simon also helped to ensure that this book reflects the great breadth of Geography.

INTRODUCTION

In writing *Studying Geography at University*, our aim is – as the subtitle suggests – to enable you to succeed in the first year of your degree in Geography. If you are studying Geography – be this Physical Geography, Human Geography or Environmental Science – or if you are studying Geography combined with any other university subject, then this book is for you.

Our approach to writing the book has been simple. First, we wanted to write down everything we wish we had known about starting a Geography degree when we were students. While we had both enjoyed studying Geography at school and college, we were both some of the first members of our family to attend university, so it turned out that we had very little idea about what to expect from our Geography degrees!

Second, we wanted to write a book informed by our experience of teaching Geography at several universities. To date, between us we have amassed nearly 30 years of experience doing this and have been lucky enough to win one or two prizes for our teaching along the way. In this time, we have also served as external examiners at other universities; been a member of the Royal Geographical Society's panel that accredits undergraduate Geography degrees; been part of courses at our own and other universities to help staff to develop into better teachers; been a 'senior tutor', overseeing the pastoral care of undergraduate Geography students; worked with a range of UK exam boards to update their GCSE and A-level Geography syllabuses; run updating and upskilling courses for schoolteachers; and written numerous academic journal articles, book chapters and reports on the topic of making a successful transition from studying school- or college-level Geography to studying degree-level Geography. All of this has given us both a broad and deep understanding of the issues associated with starting a Geography degree, which we have used to inform this book.

Third, we wanted to present the information in a short and, hopefully, easy-to-read form that is relevant to your first year. We have both been the leader of a major first-year module in Geography at Newcastle University called 'Geographical Study Skills'. Many of the topics and issues that we have encountered in leading this module – from the perspectives of both the students and the tutors who worked with us – contribute to the content of this book. Some of the information is academic and some of it is non-academic, but it all contributes to your experience of being a Geography student.

Finally, as this book was written for Geography students, we were determined that, throughout, it should include the voices of recent Geography graduates. Therefore, we are indebted to graduates of a wide range of universities, who spoke to us openly and honestly about their experience of studying Geography at university. Their insights have informed every chapter, and the book is the better for it.

How to use this book

You can engage with this book in several ways. Of course, one way is by reading it from beginning to end before your Geography degree starts. Another is by reading the chapter summaries, 'What Do Students Say?' and 'Top Tip' boxes only, as these will provide an overview of the key points. Alternatively, you can keep the book with you and read each chapter as a 'how to' guide as the first few weeks of your Geography degree unfold. Indeed, you could do all of these – for example, read it cover to cover at the start of your degree and then refer back to it as you settle into your life as a Geography student. We include a glossary that it might be useful to consult if there are terms or phrases that you are unfamiliar with or would like to know more about.

Whichever way you choose to read it, we hope this book will help you to enjoy studying Geography at university as much as we did.

PART I

WHAT TO EXPECT WHEN YOUR GEOGRAPHY DEGREE BEGINS

CHAPTER 1

ACCOMMODATION AND THE SOCIAL TRANSITION TO UNIVERSITY

Halls vs home

Our best guess is that you are a bit surprised to find that the first chapter of a book about studying Geography at university is about where you will live! To explain, we have always believed that beginning university involves both an academic transition and a social transition – and that there is a strong relationship between the two. In other words, if one aspect is not going well, you cannot be sure that the other aspect will go well, either.

When it comes to social transitions, one of the big, early decisions that you will have to make is where to live while studying for your Geography degree. Broadly, there are two choices – move into halls of residence (commonly just called 'halls') or continue to live at home. Beyond the obvious, this choice is important as it impacts on how you will make new friends, how you will interact with the university campus in the early weeks of university and, therefore, how you will go about settling into your new Geography degree course.

Most universities provide *halls of residence* that are safe, comfortable and good value. In some big university towns and cities, private companies are now also offering halls of residence. Halls are usually furnished flats, with several single bedrooms sharing a kitchen, toilet, bathroom and lounge area. Some are catered, others part-catered or self-catered. Some are mixed-sex, others are single-sex halls. While the more modern halls are en suite, there is nothing like sharing the living, cooking and washing facilities to help you to get to know other students quickly and to make new friends. Indeed, students enrolled on larger Geography degree courses often become friends with students from their halls before they make friends with people on their course.

As a result, the commonly held view is that students who live at home 'miss out' on some parts of the student experience, because they are physically detached from the campus and other students for large parts of each day. We both lived at home throughout our undergraduate degrees, and a sense of isolation and 'missing out' were not feelings that we experienced, but for some new students living at home these feelings can be very real. That said, looking back, we can now see that continuing to live at home did change our experience of being students: it was not worse, just different. For example, for students who continue to live at home, the first university friends they are likely to make are those also taking their Geography course. These are really useful friends to have, as students studying other subjects experience the transition to university differently, and so being good friends with fellow Geography students creates a more supportive sense of shared experience. However, for students who live at home, the process of making friends will inevitably be slower as they spend less time with students on their course than people who live together in halls. In the short term, this can leave the students living at home (us included) questioning themselves and what is going on. Am I suddenly unpopular? No! Do I fit in? Yes. It is just possible that you will not make as many good friends as quickly as you might expect.

TOP TIP

Remember that the transition to university is a process rather than an event, and will go on well beyond induction week. Don't put too much pressure on yourself to have made lots of good friends and to be on top of your studies in the first couple of weeks.

Getting into a routine

Usually, one area where students who live at home have an advantage over students living in halls is that they often manage to get into, and stay in, a routine sooner. Independent learning is frequently spoken about as a key part of being an effective Geography student (see Chapters 3 and 4). Our view has always been that an equally important first step is getting to grips with independent living. Everyone can really study effectively only once they have the other aspects of their lives sorted and have routines for washing, shopping, cooking, cleaning, sleeping and so on. For students living in halls, these things take a while to establish. And for some, these take so long to establish that their studies begin to suffer. This is understandable, as it is a tricky aspect of becoming a student yet one in which most universities offer no training.

WHAT DO STUDENTS SAY?

'In my first few weeks I partied so much. Everyone did it, so I felt I had to join in, even though it wasn't really me. After about a month I was broken, and I realised I couldn't keep it up. From then I went out less, did more of the things I actually wanted to do and I made some of my best university friends through societies and at work in Tesco.'

(Josie, graduated with a BA (Hons) degree in Geography)

In contrast, students who live at home tend to have fewer of these social complications. In part, this is due to logistics: their bus or train comes at a certain time each day, so they have to be ready to head into university. It is too far to travel home in the gaps between lectures, so they use this as study time in the library (see Chapter 13). The last bus home is at midnight, so the options for staying out drinking late are reduced. They still have friends from school or college living close by, so this part of their life tends to continue as normal. However, in part, their routine is also governed by practicalities, especially when their parents continue to do most of the washing, shopping, cooking and cleaning. As several aspects of their life continue as before, the social transition to university can come as less of a culture shock. The knock-on effect is that they are often able to focus on getting to grips with the academic challenges of studying Geography at university sooner than students living in halls.

After about four to six weeks, you will see that most students (whether living in halls or at home) are starting to settle into more of a routine, having fewer nights out, getting more sleep and spending more time in the library. Helpfully, for most, their body tells them it is time to slow down a little at about the same time that the first assignments are due! However, some need more help to find this balance, and this is where your tutor and university welfare support services come into play (see Chapter 7). We have both been tutors to new students for over ten years and many of our tutees have confided in us how they have struggled to settle into university. We will not tell you any more here, as (like all university tutors) our meetings with tutees are confidential. Therefore, the key message is that tutors will not be shocked by what you tell them and will not judge; they will only offer helpful support and advice.

WHAT DO STUDENTS SAY?

'I'd say that the key to surviving the first few weeks successfully is self-awareness and self-discipline: learning from your mistakes; managing excesses; beginning to figure out who you really are as a person and feeling like your life is generally on the right track towards something balanced and sustainable.'

(André, graduated with a BSc (Hons) degree in Physical Geography)

Accommodation costs

Finally, we cannot overlook the importance of cost. In most cases, it is cheaper in terms of accommodation and living costs to live at home than it is to rent a room in halls of residence. That said, it is not all one-sided, as students who live at home have to incur the costs of their daily commute, whereas those who live in halls tend to be within walking distance of campus. Relatedly, apart from the financial costs, you need to consider the time cost of where you live. We both lived about 15 miles from university and so, in rush-hour traffic, we both spent more than an hour commuting from home to campus. This meant up to a three-hour round trip to attend a one-hour lecture! And, as Human Geography students, on some days we only had one hour of teaching on our timetables; some days we did not have any (see Chapters 3 and 4). The only way to cope with this was to become very efficient in managing time and prioritising tasks. In themselves, these are important transferable skills, but they seemed alien to us, having gone to school and college much nearer to our homes. It always seemed like we had to be more efficient with our time and more organised than students who lived on campus. It was not easy, for example, to study at home and just call into the university library to pick up a book that we had forgotten to collect the day before. In fact, at times, it seemed like we were juggling too many balls – attending classes; trying to find time to study in the library; trying to find time to make and to meet friends; trying to fit in shifts at our places of work; and trying to reduce our commuting costs by studying at home on some days. Over time, this all became much easier, but it taught us that there are not just economic costs to consider when deciding whether or not to live at home, but time costs.

TOP TIP

Whether you live at home or in halls of residence, there are social challenges to be faced as a new Geography student. Be aware of these – and be self-aware enough to ensure that they don't affect your health or your studies. Over time, the pros and cons of living in halls or at home even out. Most students can successfully manage either.

Summary points

1. There are pros and cons both to living in halls of residence and to living at home while you study for your Geography degree. Make sure you weigh them all up carefully and decide what is right for you.
2. Remember that it is not a choice of one or the other: many students spend some of their degree living at home and some of it with other students (either in halls of residence or in privately rented accommodation).
3. The transition to university is a process, rather than an event, and it will go on well beyond induction week. Do not put too much pressure on yourself and give yourself time to settle in.

Key reading

Christie, C. (2007) 'Higher education and spatial (im)mobility: On-traditional students and living at home.' *Environment and Planning A*, 39(10), pp. 2445–2463.

Holdsworth, C. (2006) ' "Don't you think you're missing out, living at home?" Student experiences and residential transitions.' *Sociological Review*, 54(3), pp. 495–519.

CHAPTER 2

THE TYPES OF GEOGRAPHY YOU CAN STUDY AT UNIVERSITY

What is university Geography?

We both completed our first Geography degrees about 20 years ago, but even now can remember being completely perplexed by some of the things we were taught in the first year of our degree. There was Dr Stone, who talked endlessly about Coleoptera (beetles) and their usefulness as an indicator of biodiversity; there was Dr King, who loved regression analysis and t-tests (the sort of statistics we never expected to encounter, having dropped Maths at school as soon as we could); and Professor Phillips, who, for some reason, taught us a lot about general elections! In its own way, some of this was interesting stuff, although admittedly some was not. But how, we both asked, was this Geography? And why weren't we being taught more about Ordnance Survey (OS) maps or Central Business Districts (CBDs)? With hindsight, a better question for us to have asked would have been: what is Geography? Without knowing it, our expectations of what the first year of our Geography degree would be like had been narrowed by the types of Geography that we had encountered, and enjoyed, while at school and college.

In reality, at university there are more definitions of what Geography is, and what geographers do, than you can imagine. This arises because of the breadth of the subject matter with which university geographers engage. On its website, the Royal Geographical Society (RGS) with the Institute of British Geographers provides this definition: 'Human geography concerns the understanding of the dynamics of cultures, societies and economies, and physical geography concerns the understanding of the dynamics of physical landscapes and the environment' (RGS, 2019a). The Quality Assurance Agency for Higher Education's *Subject Benchmark*

Statement for Geography (QAA, 2019) – a document that sets out the guidelines for teaching Geography in British universities – offers a similar definition. Both are a good starting point, insofar as they hint at the variety and scope of Geography.

While the breadth and diversity of university Geography is undoubtedly a positive, as it keeps your degree interesting and varied, these definitions do beg the questions: Where does Geography as a university subject begin and end? How is studying societies and economies as part of university Geography different to studying for a degree in Sociology or Economics? How is studying the physical, chemical and biological components of the Earth as a geographer different to studying these for a degree in Physics, Chemistry or Biology? Academics interested in studying Earth forms (geomorphology), water (hydrology), climate, the atmosphere and soils are based in university Geography departments; however, many more work in Biology departments, Environmental Studies and Geology departments and elsewhere.

As undergraduates in the 1990s, we struggled to answer these questions when faced with lectures on beetles, statistics and elections. Had we, we wondered, signed up to a subject that was some sort of weird mixture of Biology, Maths and Politics? We now know that the answer, thank goodness, is no! That said, it is true that Geography at degree level can have much more open boundaries than the types of Geography taught in schools and colleges. In part, this is because exam boards have made (not always successful) attempts to iron out potential overlaps between subjects. Sometimes, this has been achieved by making arbitrary decisions, for example about whether topics such as the water cycle and the weathering of rocks should be a part of the Geography, Chemistry or Biology syllabus. In contrast, universities have rarely even attempted this sort of ironing-out process and so, at first glance, close similarities and even overlaps between topics can be apparent between different degree programmes – we are sure that the Biology department would have been teaching about beetles, the Maths department about statistics and the Politics department about elections at the same time as we were studying them as part of our Geography degrees!

TOP TIP

Don't expect school-, college- and degree-level Geography to be the same. Don't expect all Geography degree courses to be the same, either – do your research carefully online and go to university open days to find out more.

So what makes Geography different from other, similar, university subjects?

Apart from the potential for overlap with other subjects that we alluded to above, one of the things that makes Geography different is the emphasis that it places upon studying any given topic through the lens of space, place, scale and/or time. In other words, while a topic itself may not look at first glance geographical, to be included as part of your Geography degree it will be studied by you from a uniquely geographical approach. Whether that is looking at the changing spatial distribution of beetles in a local area, as an indicator of changing biodiversity, or mapping and explaining the spatial distribution of votes for various political parties across the United Kingdom, the geographical themes of space, place, scale and time should be clear to see. It is this focus that both makes university Geography diverse, as a subject, and gives it coherence.

In addition, the QAA (2019) and RGS (2019a) both highlight that geographers contribute a holistic, integrating perspective on physical and human interactions and processes. Rather than study only Human or Physical Geography, they argue that geographers are most interested in the relationship between the two and how they influence each other. In other words, geographers are most interested in how people impact on the environment and environmental processes and how these processes, in turn, impact on people. This, they argue, also makes studying Geography distinct from other, similar, university subjects. Indeed, in its definition of Geography, the QAA moves beyond the traditional division between Human and Physical Geography by including Environmental Geography (focusing upon relationships between people and the natural environment) as a third significant part of the subject (QAA, 2019). Take a look at Table 2.1 (produced by the QAA) and see how many of the types of Geography you recognise. How many types do you think you have studied before? Are there any types that you have not heard of?

Table 2.1 An indicative table of the types of Geography that you might be able to study at university.

Human Geography	Physical Geography	Environmental Geography	Technologies and methodologies
Cultural Geography	Biogeography and ecology	Environmental management	Cartography
Development Studies	Climatology	Hazard studies	Fieldwork
Economic Geography	Earth system science	Regional Geography	Geographic Information Systems (GIS)
Gender Studies	Geomorphology	Resource management	Modelling
Historical Geography	Hydrology	Rural Geography	Qualitative methods and analysis
Political Geography	Meteorology	Sustainable development	Remote sensing
Social Geography	Quaternary science	Tourism studies	Spatial analysis
Urban Geography	Soils	Transport studies	

Source: QAA (2019, p. 17).

TOP TIP

If you like both Human and Physical Geography, look for opportunities to study Environmental Geography as part of your degree. Here, the focus will be on how people impact on the environment and how these processes, in turn, impact on people.

How did you get on? If there are topics in the table that you have never heard about, do not worry at all. This is normal and to be expected. You should undoubtedly be prepared to encounter some very different and new types of Geography in the first year of your degree. For example, Social Geography (focusing on, amongst other things, issues of gender and race) and Historical Geography (looking at geographies of the past and how these influence geographies of the present) are both taught much more extensively at degree level than they are in schools and colleges. Likewise, if you have not really thought about Environmental Geography before, keep your mind open, as many students enjoy learning about both contemporary environmental issues and the diversity that this type of Geography offers.

WHAT DO STUDENTS SAY?

'It was strange, I picked Geography as a degree subject 'cos I loved Geography at school. But then I sort of steadily drifted onto new things. By the time I'd reached my final year I was writing my dissertation on how humans had caused aquatic pollution in a Scottish loch and I spent a lot of my time in a lab . . . . So, my advice is not to overthink it, just go with your gut instinct and definitely study the topics you enjoy most.'

(Charlotte, graduated with a BSc (Hons) degree in Geography)

Of course, not every new Geography student enjoys all the new types of Geography – that's just not realistic! To help with this, most degree courses have a broad, compulsory syllabus in their first year, which deliberately forces you to dip your toe into the water by offering a short taste of a wide variety of these new types of Geography. You can then begin to make decisions about which types of Geography you will pursue in the later years of your degree (also read Chapter 3 for guidance on how Geography degrees are structured).

University Geography as both a natural and a social science

It is not only the topics which vary between school-, college- and degree-level Geography; it is also the style in which they are taught that can be different. Taking a 'broad brush' view, degree-level Physical Geography is often taught in a more scientific way than at school. It is common for departments teaching Physical Geography to have Chemistry and Biology laboratories and to use some of the equipment you may well have experienced if you studied these subjects at school or college. Consequently, while studying Physical Geography means gaining specialist knowledge of the environment and its processes, it also means acquiring the tools that you need to conduct scientific studies such as the knowledge of how to operate scientific equipment in the field and lab; how to design scientific studies; how to gather and curate data; and how to present an argument in written and oral form using scientific conventions. As such, you can expect a stronger quantitative element to your studies, and hypothesis testing, scientific rigour and statistical analysis will form a key part of how Physical Geography is taught at university. As a Physical Geography student, you can also expect to spend more

WHAT DO STUDENTS SAY?

'I think I spent too long in first year trying to master the statistical techniques we were being taught and not enough time thinking about how and why geographers use them. I regret that now, as I can see how they'd have been useful in lots of ways, like in my final-year dissertation.'

(Ibrahim, graduated with a BSc (Hons) degree in Geography)

time in large computer rooms using programs like ArcView and QGIS to draw multi-coloured Geographic Information Systems (GIS) maps, or RStudio and MATLAB to analyse quantitative data. This sort of software is sufficiently new for not many teachers to be confident about using it, or their schools and colleges do not have enough money to buy the powerful computers needed to run it. (Also read Chapter 21 for guidance on studying Physical Geography at university.)

Likewise, there are differences in the approach to teaching Human Geography in universities, compared to in schools and colleges. Much like Physical Geography, there is a distinct strand of Human Geography that is both statistical and scientific in its approach. Therefore, Human Geographers can also expect to spend time in large computer rooms using programs like ArcView and QGIS to draw GIS maps. However, at the other end of the spectrum, as social scientists, Human Geography students also critically engage with theories such as Marxism, poststructuralism and postcolonialism to explore and explain the world around them. In so doing, they often value discussion, debate, argument, evaluation and criticism of ideas about the need to find the definitive, 'right' answer. Many of these are academic skills, which we will talk about later in the book, but for now it is important to note that often the marks for Human Geography assignments lie in the quality of the argument and discussion, not for giving the 'right' answer. For Geography students with a scientific mind, this focus on opinions can make Human Geography appear somewhat 'wishy-washy'. On the other hand, other Geography students can perceive the more quantitative, scientific approach to Human and Physical Geography as 'difficult' and 'not for them'. Figuring out which approach you prefer is one of your key tasks as a first-year Geography student.

WHAT DO STUDENTS SAY?

'I never imagined when I arrived at uni that I'd be studying Karl Marx as part of my Geography degree, and I didn't see any point to theory classes. Then, by about the middle of second year the penny sort of dropped and I "got" why theory was important. In the end, I was inspired enough by Marx to write my dissertation on social inequality in Liverpool.'

(Harry, graduated with a BA (Hons) degree in Geography)

Differences between universities' Geography departments

How you experience Geography at university will vary significantly between universities. For example, while all will teach both qualitative and quantitative Human Geography, the extent to which each is taught can vary significantly between universities. For instance, you are likely to find more quantitative Human Geography in universities like Bristol, Nottingham, Leeds, Manchester and Edinburgh. By contrast, while Exeter University teaches more Historical and Cultural Geography than Newcastle, Newcastle University has more modules in Economic, Political and Social Geography. The same is true of Physical Geography. For example, there are more opportunities to study oceans and oceanography at Southampton and Bangor universities than at Manchester or Northumbria universities.

Why do variations like these arise? Well, for two reasons: first, the *Subject Benchmark Statement for Geography* (QAA, 2019) allows universities greater flexibility in what is taught at degree-level than in the syllabuses used in schools and colleges; second, most university Geography departments also conduct research, and it is helpful to have several members of staff working in a similar field so that an area of expertise is built up. Consequently, the types of Geography taught in various universities tend to cluster around these areas of staff expertise, particularly in the later years of degree courses.

TOP TIP

Look at the course content of Geography degrees carefully before you apply. While most applicants worry about which degree programme is the 'best', often there is more variation in terms of the types of Geography offered by university Geography departments than there is variation in the 'quality' of their courses.

The implications of this for you, as a new Geography student, are twofold. First, before you apply for a Geography degree course through UCAS (Universities and Colleges Admissions Service), it is worth poring over the university websites to see in detail which courses are most appealing to you. Second, once your degree is underway, it is sensible to reflect throughout your first year on which types of Geography you are enjoying the most and why, initially trying to figure out whether you are mainly a Human, Physical or Environmental Geographer and then, within these broad categories, the type of Geography you would like to pursue in the next year of your degree. There is no sure-fire way to know this before you apply. Many of our students tell us at open days that they are Human Geographers and then go on to study mainly physical modules, and vice versa. Many more now see themselves as Environmental Geographers, who are interested in the relationship between Human and Physical Geography. Learning about the various types of Human, Physical and Environmental Geography, and positioning yourself within this, is all part of making a successful transition to university in the first year of your degree.

TOP TIP

Don't worry about the variety of new types of Geography you can study as part of most Geography degree courses or whether you prefer Human or Physical Geography. Very few of our students stick to the familiar topics from school or college throughout their degree. Most feel re-energised by the opportunity to study new things.

Summary points

1. There are significant differences between the types of geography that taught as part of school-, college- and degree-level Geography. Do not worry about this, as most new Geography students enjoy the chance to study something different.
2. Not all universities' Geography departments teach all types of Geography. Do not worry about this, and carry out some careful research before applying to UCAS to find the one that will suit you best.
3. Many students who think they are Human Geographers when they arrive at university evolve into Physical Geographers during their degree, and vice versa. Many more see themselves as Environmental Geographers, who are most interested in the relationship between Human and Physical Geography.

Key reading

Tate, S. & Swords J. (2013) 'Please mind the gap: Students' perspectives of the transition in academic skills between A-level and degree-level geography.' *Journal of Geography in Higher Education*, 37(2), pp. 230–240.

Ferreira, J. (2018) 'Facilitating the transition: Doing more than bridging the gap between school and university geography.' *Journal of Geography in Higher Education*, 42(3), pp. 372–383.

CHAPTER 3

DEGREE ORGANISATION AND STRUCTURE

Modules and credits

A module is a mini-course about a particular topic within Geography, such as globalisation, tectonics or climate change. Most university Geography degrees are modular, and the modules of your degree will be worth varying numbers of credits, depending on how much time you are expected to spend on learning about the topic. For example, a 10-credit module will require an average of 100 hours of learning, while a 20-credit module will require an average of 200 hours of learning. All full-time UK-based Geography degree courses are worth 120 credits per academic year, which means that they all require you to learn for around 1,200 hours per academic year.

> ### TOP TIP
>
> Each module will have its own topic, teaching staff, classes and assessments. Use your module choices to study the types of Geography that are most interesting to you and are assessed in the ways that you prefer.

First-year modules tend to be compulsory and therefore the class sizes are larger, with 200+ students at some universities. They are intended to give all students a common grounding in key geographical themes and concepts. In contrast, final-year modules tend to be more specialised. There will also be more options (sometimes called 'electives') to choose from in your final year, reflecting the fact that by this stage you will be developing into the type of geographer that you want to be. Every student has different geographical interests and chooses different combinations of modules; this naturally shares out the cohort and makes final-year class sizes smaller, with only 15 to 20 students per module at some universities.

WHAT DO STUDENTS SAY?

'My first-year lectures had nearly 300 people in them and I only knew about five of them well, so I never arrived late as I felt everyone was watching me . . . . By the end of the first semester I realised that not sitting next to my friends in lectures didn't matter so much, and it was more important to know people in my seminar and lab groups, where we'd be expected to do work together and discuss things.'

(Mia, graduated with a BA (Hons) degree in Geography)

Who will be teaching me?

Commonly, the academic staff who teach in university Geography departments are referred to as lecturers. However, in most UK universities, in reality there are five academic job titles. The majority of academic staff are either 'teaching fellows', 'lecturers', 'senior lecturers' or 'readers' (although in some universities the title 'associate professor' is used instead of 'reader'). These members of staff are properly addressed by their academic qualification: in other words, as 'Dr' if they have a PhD, or as Mr, Mrs, Miss, Ms and so on, if they do not. Traditionally, heads of departments and other senior academic leadership roles at a university are undertaken by those holding the fifth job title: a 'professorship' or 'chair'. Academic members of staff holding this title are properly addressed as 'professor'. So, while there is a hierarchy of seniority between an imaginary Dr Smith and a Professor Jones, both will be subject specialists in their area of Geography, both will have a teaching qualification and/or years of experience teaching undergraduate students, and both will teach you as part of your Geography degree.

TOP TIP

Many university Geography lecturers are happy to be called by their first name, but some aren't! Until you are sure, err on the side of caution and address academic staff using their title. For example, call us 'Professor Tate' and 'Professor Hopkins' until you've worked out that we are more than happy for students to call us Simon and Peter!

Every module will have an academic member of staff who has been designated as its module leader, sometimes called the module convenor. They could have either the title of Dr or Professor, as it makes no difference to the quality of the module which one the module leader has. The module leader has responsibility for the design and the delivery of the module, for coordinating the teaching and assessment on the module and for carrying out the administrative and quality assurance processes required by the university. Some modules that you study will be taught only by the module leader; other modules, particularly those in the first and second years of your degree, will be taught by several members of staff. On some modules, the module leader will mark all of the assignments; on other (usually larger) modules, the marking will be shared between several members of staff.

In addition, in many universities it is likely that at some point you will be taught by postgraduate students. These are students who have performed so well in their undergraduate degree that after (i.e. 'post') their graduation they decided to study for a PhD in Geography – and are therefore a long way down the road to being subject specialists (and being addressed as Dr) in their own right. Most universities insist that their postgraduate students undertake a teaching qualification before delivering any classes so that the quality of their teaching is assured. Postgraduate students are also part-time members of staff, as they are paid by Geography departments to teach seminars or support academic staff with the delivery of workshops, computer classes, lab practicals or fieldtrips. Their teaching is closely supervised by module leaders; therefore, postgraduate students rarely lead modules.

Lastly, at some point in your degree you will also encounter information technology (IT) and lab technicians, health and safety officers, professional service and administrative staff, who all play a vital role in supporting teaching and learning on Geography degree programmes.

WHAT DO STUDENTS SAY?

'At first, it seemed odd being taught GIS by a postgrad. However, I soon realised they were really knowledgeable, and they often knew in advance the bits we'd get stuck with. It was often less scary to ask them what seemed like daft questions than the module leader!'

(Stefan, graduated with a BSc (Hons) degree in Geography)

How much time will I spend each week learning about Geography?

One of the most difficult questions you can ask a Geography lecturer at an open day is 'How much contact time is there on your degree course?' In other words, 'How much time will I spend each week being taught Geography?' This is a difficult question to answer for three reasons. First, each week of your Geography degree can be very different. For example, you might have a lot of contact with academic staff during a residential fieldtrip but much less in a week of lectures and seminars. You might not want much contact with lecturers in the run up to an exam, and may prefer to be left alone to get on with your revision. Quoting any sort of average hides these important differences. Second, the question is often asked on the presumption that more contact time with lecturers means that you will learn more and therefore will be getting better 'value for money'. This is a difficult belief for us to change during a brief chat with a potential new student and their parents, to whom we are also trying to promote our degree course – although, in the next chapter, we will have a go at persuading you that learning about Geography at degree level involves a lot more than being taught! Third, it is a difficult question to answer because any plausible number that we come up with as an average amount of contact time with lecturers per week will sound low to a student who has recently completed their school or college education.

Of course, this all sounds like us avoiding the question and just makes you even more curious about the average number of contact hours that you should expect on a Geography degree. So, keeping everything above in mind, let's go with a very approximate number of somewhere in the region of 10 to 15 hours per week. Any UK-based Geography degree course should have somewhere around that figure. To understand why this figure is typical, you need to know more about how universities approach teaching and learning Geography – and that is the focus of the next chapter.

WHAT DO STUDENTS SAY?

'I remember that every open day we went to my dad was obsessed with asking lecturers how many contact hours I'd receive each week. By the middle of second year and into third year, we all wished we had fewer contact hours, so we had more time to read and write our assignments!'

(Carolyn, graduated with a BA (Hons) degree in Geography)

Summary points

1. Most university Geography degrees are modular. Modules are worth a varying number of credits, depending on how much time you are expected to spend learning about the topic.
2. All modules that are worth the same number of credits involve the same number of hours of learning. However, some modules may have more time allocated to teaching than others.
3. You should expect to be taught by a range of staff as part of your degree, from professors to postgraduate students.

CHAPTER 4

APPROACHES TO GEOGRAPHY TEACHING AND LEARNING

The differences between teaching Geography and learning about Geography

At the end of the previous chapter we said that, at a very rough estimate, you should expect to be taught for somewhere in the region of 10 to 15 hours per week. Although this figure is fairly typical of all UK-based Geography degrees, if pushed and we mention this figure at an open day, the next question that inevitably arises is that only spending 10 to 15 hours per week being taught Geography sounds like a part-time course. We would agree, if teaching were the only form of learning that took place at university. It is not and, arguably, the time spent in class is not even the most important form of learning that takes place on a Geography degree.

> ### TOP TIP
>
> More contact time means you are being taught more – but it doesn't necessarily mean you are learning more about Geography or learning it better.

Good university teaching should be inspiring. Your Geography lecturers should make you want to know more about it and about the world around you. As we'll explore more in Chapters 5 and 6, good Geography lectures should facilitate your learning by giving you some key themes, ideas and authors to think about in relation to the topic; and good Geography seminars and workshops should give you the opportunity to discuss, debate and apply these in more detail. What good Geography teaching at university will not do is to try to tell you everything that there is to know about a topic or what to think about it. That deeper learning is up to you. Sometimes, that will mean spending time in the university library, reading more about the topic. Sometimes, it will mean learning-by-doing, such as spending time in a computer room learning more about ArcGIS or QGIS by drawing and interpreting maps. Sometimes, it will mean a self-directed fieldtrip around the local area, trying to work out the impact of a concept discussed in a lecture on the local environment. This is all really important learning, and it can take up a lot of time – but it is not teaching, and so is not included in the figure of 10 to 15 hours per week.

WHAT DO STUDENTS SAY?

'At my old school, the focus was on being taught, with some independent study time. At university, I suddenly found that the balance had changed and that wasn't something I was really expecting …. . If I'd just remembered everything I was taught in these classes and not done much extra reading, I'd have probably just about scraped a pass in my degree, but not much more.'

(Kadeeja, graduated with a BA (Hons) degree in Geography)

So, clearly, there is a balance to be struck. Too little teaching and, as a student, you would not know where to start learning about a topic; however, too much teaching leads to 'spoon-feeding' information, which students remember and regurgitate in their assignments and this, in turn, leads to a fairly narrow and uncritical approach to learning about Geography. Getting this balance right is the job of every module leader and, in many cases, where the balance lies depends on the topic that they are teaching. For example, if you study a GIS module or a hydrology module as part of your degree, in addition to lectures you will most likely be required to spend time in a computer room or a laboratory. These modules will need more teaching hours, as it would be unfair at first to expect you to figure out intuitively which buttons to press on the keyboard to make the software work. Equally, it would be unsafe to ask you to work in a lab without either training or supervision. In contrast, if, for example, you study a module on everyday geographies or the geographies of film as part of your degree, you should expect to spend more time without staff supervision. Instead, you might be expected to learn by thinking about how the concepts that you have been introduced to in lectures and seminars relate to either the movies that you have been asked to watch or people's day-to-day lives that you have been asked to observe ethnographically.

We said in Chapter 3 that you will spend the same amount of time learning on every module with the same credit weighting. We can now see that, depending on the various needs of topics, you should expect some variation in how much time you spend being taught on modules of the same credit weighting. So, the diversity of Geography as a university subject means that you can choose not only to study the topics that are of most interest to you, but to learn in the ways that suit you best.

Managing your independent learning time

Whichever combination of modules you choose, independent learning will be a key part of studying for your Geography degree. It will challenge your motivation and dedication while developing your organisational and time management skills. Being an independent learner means being an active learner, taking responsibility for your workload, commitments and deadlines. Consequently, one of the most important academic challenges that you need to tackle very early in your degree is getting into a good independent learning routine. Once mastered, it will be the key to success in your Geography degree. Here are some things to think about:

1. One starting point is to try to work out how you study best. You can only do this by trial and error. For example, do you prefer to work late at night or early in the morning? Are you a perfectionist who will spend far too long on even the smallest of tasks? Do you like to leave things to the last minute, needing the stress of a deadline to make you focus? Or do you need to start tasks early, because you find the stress of a fast-approaching deadline makes it difficult for you to concentrate? Consider the extent to which you can realistically work with your personal preferences and the extent to which they need to be changed because they will not help you to produce your best work.

2. The second thing to figure out is where you study best. Is it in your bedroom? In the university library? In a different on-campus study space, such as a student common room? Again, this comes down to personal preference and, instead of expecting to start university as the 'perfect' Geography student, use the first year of your degree to experiment – trial and error is the key.

WHAT DO STUDENTS SAY?

'I found the amount of independent study expected overwhelming at first and I found it really difficult that most staff didn't check whether you'd done any. It was obvious who in the class was working hard from the assessment marks but, without someone checking up weekly like a teacher, it was hard at first to be self-motivated. That was definitely a new skill I had to quickly learn!'

(Serhat, graduated with a BSc (Hons) degree in Geography)

3. Remember that independent learning does not necessarily mean working on your own. Working with other Geography students, encouraging each other and talking through difficulties may be the most effective way of working independently. Explaining a concept or idea to another Geography student is a very good way to make sure you understand it. If you have a problem, explaining it to other Geography students can help to clarify the issues for you. You can also share books and other learning resources. These are all reasons why many Geography students choose to be part of reading groups and study groups with other Geography students. It is also one of the reasons why many university libraries now have 'social study' spaces. Just make sure that working in a group does not mean that you end up intentionally or unintentionally plagiarising other students' work (see Chapter 16).

4. At the end of each week, ask yourself honestly 'How many hours did I spend on independent study, and was it enough?' The answer to the question of whether or not you have done 'enough' will depend upon the type of week it

has been. For example, what constitutes enough hours of personal study will be much higher during a revision week than on a residential fieldtrip week. That said, assuming that your university year lasts around 28 weeks and you are taught 10 to 15 hours per week, as a very rough guide an average of 20 to 25 hours per week of independent study is in line with the idea of 1,200 hours of learning per year. Doing a degree is more or less equivalent to doing a full-time job!

5. Ask yourself whether the time you spent on independent study was time well spent. Did you work effectively? You may have spent all day in the library, but were you studying all day? Did you lose your concentration? Rather than getting on with them, were you putting off difficult tasks by procrastinating?

WHAT DO STUDENTS SAY?

'I found routine really important. I'd look for days on my timetable where I had big gaps between lectures, and I'd block out some of that time to go to the library and read. I'd force myself not to miss it, in the same way I wouldn't miss a lecture. I found doing chunks of reading across the week really helped. In the same way, I'd also block out times during the week to go to the gym and I'd treat that as protected time too, as it helped to structure my week.'

(Sam, graduated with a BA (Hons) degree in Geography)

Asking for help

Of course, this does not mean that your only contact time with Geography lecturers will be while being taught in a lecture, seminar, workshop, practical, lab class or on a fieldtrip. Help and support with your degree will be available if you need it. However, as a Geography student at university, you will need to be proactive, recognising when you need help with a module and asking for it.

TOP TIP

Don't wait to be asked if you need help. If you aren't asking for help with your studies, academic staff will assume you don't need it!

The most common ways of asking for help are either to send your lecturer an email or to call into their office. Most lecturers either have an open-door policy or set aside some time in their office each week for students to drop in to ask questions about their module or to have a chat. Each member of staff may favour a different approach, and you will soon pick up which they prefer. Equally, you will soon get used to talking to your lecturers one-to-one about your questions. Most lecturers really enjoy answering students' questions, as engaging with students in this way helps us to figure out how clearly we have explained things in our classes. It is sometimes really hard for lecturers to work this out in a lecture environment, so never feel embarrassed about asking for more clarification or explanation afterwards. If nothing else, it helps you and gives us another opportunity to talk about a topic that we are passionate about!

This also means that you have to direct your questions to the right lecturer, and not just the one you can get hold of first. Usually, the right member of staff to speak to will be the one who taught the class to which your question relates. Alternatively, you could choose to speak to the module leader, if this is a different person. You should not expect any Geography lecturer to be able to answer any question about any module – university teaching is too specialised for that to be realistic.

Why is this approach to teaching and learning worth it?

Apart from giving students a better understanding of Geography as a subject and the world around them, the way in which students learn Geography at university leaves them with a broad range of transferable skills, and this makes Geography graduates very employable. The list includes being able to analyse, evaluate and interpret qualitative and quantitative information, adapting and coping with change, leading and deciding, innovation and creativity, interacting with others, organising time and delivering results, being self-motivated, as well as being highly literate and numerate. These are the skills that graduate employers want and it makes students graduating with a degree in Geography very employable.

Many Geography graduates work in professions where they use their geographical knowledge in sectors such as housing and real estate, flood/environmental prevention, transport planning, logistics, sustainable development, community work, teaching and surveying. However, geographers are well placed to enter professions as diverse as accountancy, marketing, retail, human resources, the civil service, administration and data analytics. Many geographers go on to become self-employed or work for charities in a variety of positions. In fulfilling these roles successfully, they rely on the transferable skills that they acquired while studying for their degree.

TOP TIP

About 70% of graduate jobs are open to graduates from any subject and the range of transferable skills that Geography students possess makes them highly employable in a range of professions. Ask your university's careers advisors for more information about the types of jobs that geographers can do.

Summary points

1. Whichever university you study Geography at, expect there to be more independent learning required than at school or college. Self-motivation and managing your time are therefore crucial skills if you are to graduate with a good Geography degree.
2. Be confident and learn to ask for help when you need it.
3. Because of the transferable skills that they possess, Geography graduates are very employable in an extremely broad range of graduate-level jobs.

Key reading

McMillan, K. & Weyers, J. (2009) *The Smarter Study Skills Companion*. Harlow: Pearson Education (chapters 6, 8 & 15).

The Making Geography Work for You website helpfully illustrates the transferable skills that you will acquire through the completion of various tasks as part of your Geography degree http://makingGeographywork.ncl.ac.uk.

CHAPTER 5

GETTING THE MOST OUT OF LECTURES

What are lectures and why are they important?

Lectures are likely to form a key component of your Geography degree, and some of your modules may be delivered entirely through lectures. While some modules will include other types of teaching (such as seminars, workshops, field exercises or lab-based activities), the chances are that lectures will provide the foundation for most modules. In your first year, some of your lecture classes may be very large, as students studying for other degrees will be attending the same lecture. This arises because the topics covered by first-year modules tend to be quite broad, and the boundaries between Geography and other subjects can be quite blurred. Consequently, the same module may be part of several degree programmes that your university offers, such as both Geography and Environmental Science. As you proceed through your degree, the lectures will tend to be more specialised, smaller and include only those students studying Geography or a joint degree (such as Geography and Planning). By your final year, there will probably be very few students from other subject areas in your Geography lectures (see Chapter 3 for guidance on modules and how your degree is structured).

Lectures normally last an hour, but some modules may have time scheduled for a two-hour lecture slot. Subject specialists – who may have researched and published widely on the topic that they are lecturing about – deliver lectures. In Physical Geography, the lectures often include the delivery of information about scientific processes associated with the physical environment; whereas in Human

Geography the lectures tend to focus on outlining the key issues, debates and arguments associated with specific geographical issues. In both Human and Physical Geography, lectures rarely tell you everything that you need to know about a topic, therefore independent learning is needed to supplement the information given (see Chapter 4). To help you to get the most out of lectures, it is useful for us to consider what you do before and after the lecture, and not just what you do when you are in the lecture theatre.

Before the lecture starts

Before the lecture, you should familiarise yourself with the location, time and topic of the lecture, and engage in some background reading. Your module handbook may include a summary of the lecture or a set of learning objectives that it is useful to read over in preparation. Many modules include recommended reading lists, and there may be a chapter or list of articles – sometimes referred to as 'papers' – linked to the lecture that you are preparing for, which you should read through and make notes on (see Chapters 11 and 12). It is useful to consider where the lecture is in the scheme of the module as a whole, and if there are related lectures or this is a standalone lecture on a specific topic. You should also consider if there are seminars, workshops or field classes connected with the lecture and how the lecture links to specific assessment requirements. In preparation for the lecture, you will also want to ensure that you have everything you need to get the most out of the lecture. This includes paper, pens, highlighters, a laptop, your module guide and any notes from related lectures or reading. Possible distractions are worth considering here – consider turning off your phone (or turning it to silent) and closing down Facebook, Instagram or other social media on your phone or laptop (unless the lecturer specifically requests you to use these as part of the lecture).

TOP TIP

In preparation for each lecture, read a key book chapter or journal article related to the topic of the lecture.

In addition to preparing yourself academically, it is important to prepare physically and psychologically for each lecture. You want to be alert and ready to actively listen and take notes, so it is important to ensure that you have had plenty of sleep so that you are awake and alert enough to absorb the lecture content. You should also consider if you have had enough to eat and drink, especially if it is early in the morning, around lunchtime or in the late afternoon. Many universities do not stop teaching for a designated lunchbreak, so you want to make sure that you have enough energy to keep focused and not lose concentration during a midday lecture – an hour is a long time to concentrate fully. It is also useful to consider what you should take into the lecture theatre with you, such as a (refillable) water bottle, to keep hydrated.

During the lecture

Having prepared yourself academically, physically and psychologically for the lecture, you are now in a strong position to get the most out of it. Assuming that you are on time and prepared, an important decision upon entering the lecture theatre is where you should sit, and whether it is better to sit with friends or on your own. You are more likely to maintain your focus if you sit near the front or the middle of the lecture theatre, as you will be closer to the lecturer. It can be easier to be distracted if you sit near the back row of a large lecture theatre. It can also be difficult to hear clearly what the lecturer is saying. Although sitting with friends may add to the list of possible distractions, there are added benefits, too, as you could help each other to ensure that you do not miss any key points.

Just before the lecture starts, it is useful to write down in your notes key information, such as the lecturer's name, the module title and code, the title of the lecture and the date. This will help to ensure that you have a coherent set of organised notes as the module progresses. Most lectures will have a structure (much like that of an essay – see Chapter 14), where the topic or issue is first introduced and the aims of the lecture set out, followed by the main body of the lecture where the key issues are explored, with examples, which is followed by a conclusion in which the key themes and summary points are identified. Even if the lecture is not structured explicitly like this, it is worth listening carefully so that you can identify points that are introductory, as opposed to those that are a substantive part of the lecture.

> ## TOP TIP
> There is no single model for delivering a lecture, and your lecturers will vary in their approach. In order to get the most out of lectures, you need to be able to adapt to the styles of your lecturers.

A key challenge to getting the most out of lectures is about listening carefully (see Chapter 10) and taking notes effectively (see Chapter 11). If you can master both of these skills, you will get a lot more out of lectures. Remember that a lecture involves a relationship between the lecturer and the students; different lecturers have different presentation styles and approaches to lecturing. Try not to jump to conclusions about how good or bad a lecturer is, based only on their specific style or other personal characteristics. Although we all enjoy lecturers who adopt an entertaining or techno-wizard approach, the lectures by those who at first appear to 'ramble' or 'fidget' may be more useful, and you may find that you have better notes or a more in-depth understanding of the issues following these lectures. Also, try to avoid jumping to discriminatory assumptions about a lecturer, based on their gender, sexuality, race, age or nationality. For example, female lecturers

Table 5.1 Lecturing styles and some tips for handling them.

Lecturing style	Tips for handling it
The entertainer – a lecturer who is humorous, witty and engaging. Such lecturers may use a range of tactics to entertain you – for example, we know of lecturers who dress up, sing or include humorous slides in the middle of an otherwise standard lecture.	Listen carefully for key points, phrases or points of argument and keep notes about the lecture structure. Consider whether the humour is helping you to remember things or is distracting you from the topic at hand.
The droner – a lecturer who monotonously delivers material without much expression.	Listen carefully for key points, phrases or points of argument and keep notes about the lecture's structure. Consider the ways in which the slides or handouts point to key issues or important points, if you are not able to gauge this from the lecturer.
The rambler – a lecturer who appears to wander from the point and goes off on a tangent.	Keep focused and take notes, as what appears to be a random point may be useful to note as the lecture progresses. Listen out for key points or signposts, and devise a way of noting these down.
The mumbler – a lecturer who does not project their voice well, particularly a problem in large lecture theatres.	Sit nearer the front of the lecture theatre so you can hear more clearly. If this does not work, you could speak to the lecturer at the end of the lecture to express your concerns, or to the student representative for your class or year group.
The fidget – a lecturer who either fidgets with a pen, paces up and down or waves their arms about a lot during the lecture.	Listen carefully to the content and the issues being explored while taking notes. Consider whether the fidgeting or pacing of the lecturer is indicative of key points (for example, a lecturer may raise their hands in a particular way when making a significant point) or whether it is just a habit.
The techno-wizard – a lecturer who uses diverse forms of technology during the lecture. Such lecturers may be motivated more by the use of diverse technologies than a traditional lecture format.	Maintain focus on the content and issues being explored rather than on the snazzy technology being used, while considering if there are useful points to learn about technological innovation, too.
The egotist – a lecturer who focuses on their own specific work or promotes their own research, with little consideration of different viewpoints.	Listen carefully and take notes. After the lecture, check your reading list and the library for the work of other scholars who have published on the same or similar topics.

Source: MacMillan & Weyers (2012, pp. 93–94).

tend to receive more unjustly critical feedback (particularly from male students), while older men tend to receive positive feedback. Hopefully, we can all agree that this is not fair or right.

The challenges experienced by students in lectures tend to involve either losing concentration or being left behind. Maintaining concentration during the lecture may be easier if you use techniques to maintain your focus, such as drawing diagrams related to the content, using colour or putting things into your own words rather than simply copying from the lecturer's slides (see also Chapter 11 for more guidance). Some students complain about a feeling of being left behind in lectures, as the pace is fast and they feel that they are struggling to keep up. Being prepared for the lecture (as suggested above) can minimise the likelihood of this happening. If you do find that you are struggling to keep up, it may be useful to share notes with a friend afterwards. You could also look at the lecture slides or handouts provided or speak to the lecturer at the end to ask them about specific issues that you missed or did not fully understand.

WHAT DO STUDENTS SAY?

'In first year I could never keep up with the lecturers, because I was trying to write down everything they said. Eventually, I realised this didn't really help me very much, so I started to read a couple of the key readings beforehand. I found it really helped me to understand the lecturer more clearly and I started to realise that in the lecture I only needed to write down new points, things that the lecturer said that weren't on the slides, and links I could make between what was said and the stuff I'd read. Most of the lecturers gave us copies of their slides anyway, so it was a bit pointless copying these out!'

(Michaela, graduated with a BSc (Hons) degree in Geography)

After the lecture

The worst thing you can do after a lecture is to pack away your notes and never look at them again! Revisiting your notes as soon as possible after the lecture is useful so that you can clarify any points that you had missed or which you now have difficulty in re-reading. Talking with friends about the lecture and comparing notes may be helpful to your understanding and to ensuring that you grasped all the material. Also, consider how each lecture fits into the module as a whole and if there are connections between lectures that are worthy of consideration.

After a lecture, it is useful to visit the library to read up further about the topic. Did the lecturer identify key readings during the lecture that would be useful to follow up on? Was a significant debate or contested issue referred to, which you could find out more about? Could you read up on this to find out more about some of the ways in which the topic or issue has been presented and challenged? Perhaps there is a key scholar whose work was mentioned a lot during the lecture? It may therefore be useful to find out more about this person, their publications and the type of research that they have been involved in (see Chapters 12 and 13).

> **TOP TIP**
>
> Revisit your notes as soon as possible after the lecture to add any points that you missed or to clarify any points that are unclear or difficult to read.

Summary points

1. To get the most out of lectures, consider how you prepare academically, physically and psychologically so that you are in the best possible position to actively listen and take notes.
2. Do not be daunted by very large class sizes – at the start of each lecture, find a seat in the best place for you.
3. After the lecture, read through your notes to fill in any gaps and to ensure that you understand them visit the library to engage in further reading about the lecture topic and consider discussing the lecture with friends to clarify your understanding of the key points.

CHAPTER 6

GETTING THE MOST OUT OF SEMINARS AND TUTORIALS

What are seminars and tutorials?

There tends to be a close relationship between the material discussed in lectures, seminars and tutorials. While the lectures give you a broad overview of a topic, the seminars and tutorials tend to focus on specific issues or case studies relevant to the course material. Seminars involve a group meeting of around 10 to 15 students with a seminar leader; however, the number of students may be higher or lower than this, depending on the specific practices of the department or university in which you are studying. In some universities, the terms 'seminar' and 'tutorial' are used interchangeably; in others, tutorials refer to either individual meetings between a tutor and a student to talk about pastoral issues (see Chapter 7), or a very small group meeting of between three and four students and a lecturer to discuss an academic topic.

TOP TIP

It is important to understand early on how your Geography department defines its 'tutorials' so you know what to expect from every session that you attend.

Some of the modules that you study at university may have weekly seminars or tutorials. Others may have them only every two or three weeks, and others still may organise seminars or tutorials only at specific points in relation to the material being discussed (see Chapter 4 for a discussion on degree modules). This means that your timetable will vary from week to week, and you need to be careful not to miss classes. That said, seminars and tutorials tend to be a more regular feature of Human Geography modules than Physical Geography modules, where there tends to be more laboratory-based classes and practical exercises – you may even have a Human Geography module that is taught solely through seminars and tutorials. As you proceed through your Geography degree, the level of difficulty of seminars and tutorials will increase, as will the expectations of you.

Your seminar leader (or tutor) may be the same person as you met in lectures. However, it could be another member of the academic staff, a researcher, a technician or a specialist in a specific field. In some universities, postgraduate students are paid to lead seminars. These are students who performed so well in their undergraduate degree that they are studying for a PhD in Geography, and are therefore a long way down the road towards being subject specialists in their own right (see Chapter 3 for a discussion on the range of staff who will teach you as part of your degree).

Before every seminar or tutorial

For most modules, there will be several groups for seminars and/or tutorials, so it is important to note the specific group that you have been allocated to, as this may not necessarily be the same group as your friend is in. Because of timetabling issues, it might even be a different group for each module that you are studying.

Seminars and tutorials take many forms, and often they differ between modules and across the year. Figure 6.1 outlines some of the types of seminars and tutorials

- Discussion of a journal article or other reading

- Giving a presentation (individually or as a group)

- Mapping out issues

- Designing a poster

- Solving a problem or contentious issue

- Contributing to a debate

- Assessment-focused (e.g. essay planning)

- Giving feedback

Figure 6.1 Types of seminar or tutorial.

that you could participate in at university. Before a seminar or tutorial, it is crucial that you know from the list what form it will take and that you prepare by completing the tasks requested of you, so that you can arrive as organised as possible. This preparation could include a host of assorted activities: completing a reading; finding relevant sources online or in the library; undertaking groupwork; solving a problem or issue; writing up a presentation; or preparing to make a case or argument about a specific issue. These are just some examples – it is important to listen carefully to what is being asked of you for each seminar or tutorial and to prepare accordingly.

Also, consider in advance how your preparation fits in with the rest of the module's material. Does it connect to a specific lecture, or to a topic discussed at a previous seminar or tutorial? Consider making a note of any questions or issues that you are unclear about, as these may be addressed in the session. You are more likely to succeed at university if you turn up at seminars and tutorials fully prepared to participate. Having a basic understanding of the topic or issue to be explored will mean that you are more able to participate in the discussion and to understand the issues.

In advance of each seminar or tutorial, it can be useful to look over your lecture notes, as well as notes from previous seminars, especially if these are closely connected to the topic that you are preparing for. Doing this might help you to predict what will come up, so you can prepare any questions that you need to ask the seminar leader or tutor during the session.

TOP TIP

To enhance your understanding of the material, complete any background reading and/or research before your seminar or tutorial. Take notes about it and bring them with you to the session.

During every seminar or tutorial

You will get more out of seminars and tutorials if you participate fully in them. You might be nervous about speaking at first, but the point of seminars and tutorials is not to have all the right answers or to have a fully informed account of the topic. It is likely that many in your group will be nervous, too. If you are concerned about speaking up, it can be useful to try to speak first – this can help to boost your confidence. Seminars and tutorials are about learning, and about helping each other to learn, in a small-group context.

Engaging in discussion and listening to the points being made are useful ways of learning more about a topic. Seminars and tutorials present a useful opportunity for you to test out ideas and to develop your debating and discussion skills (see Chapter 17). However, when discussing ideas, even if you do not agree with them, remember to be respectful to the other members of your group (see Chapter 10 on effective listening).

During a seminar or tutorial, it can be useful to consider how you are developing your non-verbal communication skills, as communication is not only about speaking. This is partly about nodding your head in agreement or making quick comments, like 'I agree', in order to demonstrate your affinity with the person speaking. As you may already know, it can be reassuring to have someone nodding their head while you are speaking. Observing the group is important, as this helps you to assess if people are in general agreement or if there are divisions in the group, or if someone appears unclear about what is being said.

WHAT DO STUDENTS SAY?

'I was really nervous going to my first seminar at university, but it turned out that the people I met in my first group became some of my best friends at university; we used to meet up regularly outside of the seminar group and go to lectures together. Most of us ended up doing the same field trip later in our degree.'

(Tadgh, graduated with a BA (Hons) degree in Geography and Planning)

Seminars and tutorials are the ideal time for you to ask questions about the material covered in the module, about the module overall or about its assessment. Such opportunities may not happen often in lectures, particularly with large classes. If you do not want to put your question during a seminar or tutorial, you could speak to your seminar leader or tutor at the end of the session. This helps to build a relationship with them and can be a useful way of addressing any concerns that you might have. As you progress with your modules, you could suggest topics for inclusion in seminars and tutorials – this could be an issue that you are interested in, want to learn more about or need to clarify some misunderstandings about. Sometimes, you may be asked to lead a seminar or tutorial.

It can be useful to try to take some notes during a seminar or tutorial; however, this is not always possible, and it may be best to wait until the seminar is over and write your notes then. It may also be useful to consider annotating your own notes (that you wrote beforehand) during the seminar, as this may help to build up your knowledge and understanding (see Chapter 11 on effective note-taking).

After every seminar or tutorial

Many students tell us that they leave the seminar or tutorial room with their notes in their folder and do not look at them again until they are revising for their exam or completing an assignment on a related topic. Doing this means that they then find under-developed points or issues that they have not necessarily taken full notes about. Given the time that has passed since the seminar or tutorial, they are often unable to remember the specifics, and so have put themselves at a disadvantage.

In order to aid your learning, it is best to debrief yourself about the seminar or tutorial as soon as possible after it happens, as you will be more able to remember specific details at this point than you will be if you wait a few weeks. Reviewing or debriefing does not need to take a long time. It could simply be 10 or 15 minutes to read over your notes, making any additions or amendments as you go and following up on any links suggested by your seminar leader, tutor or other students.

You could debrief on your own, in the library or on the bus or train home. You could also debrief with friends in your group. Perhaps you could continue discussing the topic of a seminar or tutorial as you walk to the next class or perhaps you could go for a coffee with some of your friends and use this time to reflect on the material, how you feel about it and anything that you did not understand or were unclear about.

As well as debriefing, either on your own or with your course mates, you could consider following up on any points that had been raised during discussion. Perhaps your seminar leader or tutor pointed to a key text of an academic who has written about the topic that you were discussing in the seminar, so a visit to the library would be useful. Maybe a key issue or debate was raised that is new to you and you want to follow up on this to learn more about it. If there is a specific issue that you remain unclear about, you could also contact your seminar leader or tutor or keep a note of the question to ask them next time you see them (see Chapter 4 for more guidance on the importance of independent learning).

Summary points

1. Seminars and tutorials are likely to be a regular feature of your degree. They tend to involve a small group of students (normally up to 15) meeting with a seminar leader (or tutor) to explore a specific topic that is outlined in advance of the meeting.
2. You will get more out of seminars and tutorials if you prepare for them ahead of time by undertaking the specific tasks required of you, such as doing the reading or preparing a presentation.
3. There are many types and formats of seminars and tutorials, each of which will require a different form of preparation.
4. During a seminar or tutorial, engage in discussion, ask questions and participate as fully as you can. This enables you to maximise the benefits that you receive in terms of learning, knowledge and understanding.

CHAPTER 7

YOUR TUTOR AND OTHER SOURCES OF SUPPORT

What is a university tutor?

Tutors have different names at different universities. Sometimes they are called pastoral tutors; sometimes personal tutors or academic tutors; and, at some institutions, they are referred to as personal academic tutors (PATs). Whatever their job title, your tutor is a very important person in your university experience. Most likely, your tutor will also be a lecturer in Geography at your university. They may even teach you on some first-year modules. However, when you have a tutor meeting, their role changes from being a subject specialist to being someone who will support your broader academic and personal development. They will be your key point of contact in the university.

Your tutor will get in touch with you at the beginning of your Geography degree to arrange a meeting. While talking to a member of academic staff on a one-to-one basis may seem daunting at first, make sure that you attend this meeting, as getting to know your tutor well is important. To help with this, most universities assign each member of staff only a few students as their tutees. So, even though the number of students on your course may be quite high, you should expect your tutor to know your name and say 'hello' when you pass by them on campus.

The early meetings with your tutor will probably be informal chats, focused upon supporting your transition into university and into your Geography degree. You should expect questions about whether you have settled into university and made friends, about which modules you are enjoying most and whether you see yourself as mainly a Human, Physical or Environmental Geographer (see Chapter 2 for a discussion of the 'types' of geography and the difference between them). You should not worry about being quizzed on your subject knowledge, as that does not fall within most tutors' job descriptions! Rather, it is your module leaders who will check, using various types of assessments, that you understand the academic content of your degree, and it is they who will support you with the specific questions that you have about their module.

> ## TOP TIP
>
> Don't miss tutor meetings, and stay in close contact with your tutor throughout your degree.

As the first year progresses, your tutor will have access to all your marks and (at some universities) to your attendance records. Meetings with your tutor will evolve into reviews of your marks and general progress, making sure that you are on track academically. Your tutor can help you to understand the feedback on your assignments, and will offer suggestions about how you can improve. You can ask your tutor for advice about module choices and other study options (e.g. placements or studying abroad for part of your degree). As your degree progresses further, your tutor will encourage you to make plans for your career or for further study. When it is time to apply for part-time or graduate-level jobs, or even for master's-level courses, they will write references for you. The better your tutor knows you, the more detailed these references will be. It is hard to write a reference for a tutee whom you have met only a couple of times, and this is another reason to stay in close contact with your tutor throughout your degree.

Dealing with the unexpected

There will be several scheduled meetings with your tutor each year. However, do not wait for your next scheduled meeting if you are experiencing either academic or personal difficulties – email, text or phone your tutor to ask for an appointment or visit them during their office hours. An important part of your tutor's role is to listen to any concerns you have and to offer support and advice. We are both tutors and we really enjoy this aspect of the job. Our message to our tutees has always been that if an issue is concerning them, it is probably impacting on their ability to study to the best of their ability, and so we are here to help. However, we can help only when tutees tell us what their problems are, and that is another reason to build up good rapport with your tutor from the start of your degree – it is easier to tell your tutor about a problem that is concerning you if you already know them and they know you.

WHAT DO STUDENTS SAY?

'I found my first tutor a bit difficult to talk to. I'm sure she was a lovely person, but we just didn't click. I asked if I could change tutor, and my second one was great and helped me a lot when I told her I hadn't been feeling myself. She advised me to speak to some people and, luckily, I was diagnosed with anxiety and depression in time to get the help I needed.'

(Jake, graduated with a BSc (Hons) degree in Physical Geography)

Sometimes the concerns that our new tutees mention to us are relatively small and are easy to deal with – like not being able to find a building on campus or not having wellies for the induction fieldtrip. At other times, tutees knock on our office doors to tell us about issues that are much more serious – such as homesickness or concerns about their mental health; to tell us they have suffered a family bereavement; to tell us about an unplanned pregnancy; or that they are struggling with their LBGT+ identity (lesbian, bisexual, gay, trans, queer, questioning, intersex). Our approach as tutors is typical of most university tutors – we listen: we do not judge. And (if the student asks for it) we try to offer advice. Sometimes, tutees just want to notify us of something that is affecting them and that may impact their studies in future, even though they are dealing with it. At other times, tutees ask us for advice or want to talk through various options. Importantly, this does not mean telling our tutees what to do. In that sense, the role of a tutor is different from that of a parent or guardian and, in any case, at 18 years or older our tutees are adults, not children. Sometimes our tutees have also told their parents about the issue that is concerning them, and other times they have not. Either is fine, as our relationship is with our tutees, not with their parents or guardians – even those people are paying their tuition fees. In all but the rarest of circumstances, whatever our tutees tell us remains confidential.

WHAT DO STUDENTS SAY?

'My Geography degree was going great, then in the January of the second year my granny died in Poland. It was a total shock and I didn't know what to do. I went to see my tutor and immediately burst into tears in his office! I was in the middle of exam revision, but he was great and arranged for me to defer my exams till the summer and organised an appointment for me to speak to a university counsellor when I came back to the UK after the funeral.'

(Agnieszka, graduated with a BA (Hons) degree in Geography)

Engaging with other support networks

Of course, you should not expect your tutor to know the answer to every question! While it is relatively easy for tutors to offer advice about academic issues, to get the help you will need with other issues, from time to time your tutor will need to refer you onto someone in the university who has more specialist training. Do not mistake this for your tutor passing you and your problem off onto someone else, as an important part of the training that all tutors receive is to recognise when they need to signpost their tutees to someone else. For example, most universities and student unions have a range of trained counsellors and therapists, personal finance and student loan experts, housing experts, careers advisors, chaplains, and so on. Some universities also train final-year students to act as 'peer mentors' or 'buddies' to new students, advising on things like settling into halls of residence, making friends and living in a new city. While you can access all of these sources of advice directly yourself, your tutor can act as an important link between you and them.

A lot of Geography departments have a member of staff employed as a 'senior tutor' or as a 'student experience coordinator', who is responsible for coordinating personal tutoring and related student support. Your tutor might refer you to this person, or you might choose to visit them yourself, if a personal problem or issue could hinder your academic progress. For example, your assignment deadlines might be tough to meet if you are hit by unexpected circumstances. Senior tutors and student experience coordinators are well placed to discuss the process for requesting extensions to deadlines or for deferring exams. They can also advise about your university's processes for applying for mitigating circumstances to be taken into account when the university's Geography Board of Examiners decides your final marks.

TOP TIP

Your tutor should be able to respond to your emails or to make an appointment to see you in a reasonable timeframe. However, tutors aren't on call 24/7. For example, many Geography lecturers also undertake fieldwork, and so are not in their office every day. Many staff do not check their emails outside of working hours.

Supporting students with neurodiversity, physical and mental health conditions and disabilities

All students have the potential to do well in their Geography degree, even though many will arrive at university with neurodiversity, various physical and mental health conditions or with a disability already diagnosed. The most common form of neurodiversity amongst Geography students is a specific learning difference, such as dyslexia, dyspraxia or dyscalculia (although this is by no means a full list). Each year, between 10% and 15% of our graduating students will have been diagnosed with either dyslexia or a mental health condition such as depression or anxiety. Most manage their conditions effectively and will graduate successfully with a Geography degree. Likewise, we both remember teaching John, who attended classes in his wheelchair, accompanied by his full-time carer. We also remember Kate, who was born with achondroplasia. Both took part in university life fully, including Geography fieldtrips, and graduated with 2.1 degrees.

TOP TIP

Universities make offers based on each applicant's ability to meet the entry requirements, their enthusiasm for Geography and their references. It would be unlawful for universities to refuse you a place or treat you less favourably because of your study support needs – they have been covered by the Disability Discrimination Act since 2001 and the Equality Act since 2010.

It is not compulsory to provide details on your UCAS application about your neurodiversity, physical and mental health conditions and disabilities. You can choose to provide this information later, if you prefer. However, our advice for new Geography students with study support needs who are arriving at university is that, in order to make a smooth transition, an early talk with university staff is key. Universities have well-developed systems and procedures for admitting students with neurodiversity, physical and mental health conditions and disabilities and for making sure that they progress well in their studies. However, to do this, universities have to understand the additional support that you will need, and it helps them if you provide this information as early as possible in the application process. If you do this, they can make sure that everything is in place and that you will have the best start to your Geography course.

TOP TIP

If you are arriving at university with paperwork from a clinician or educational psychologist that diagnoses your needs and recommends what study support you will need, pass it to the appropriate person at an early stage. Your tutor will be able to point you in the direction of the specially trained staff who can review this sort of paperwork for you.

As your first few weeks at university progress, it is possible that the study support that worked well for you in school or college will not seem to be working as well for you at university. If this is the case, ask for it to be reviewed. As the demands of school, college and university are all very different, it does not always follow that what worked well in the past will work well throughout your degree. Again, talk to your tutor and they will be able to point you in the direction of the specially trained university staff who can review your study support needs for you.

Equally, it is never too late, and learning support can be arranged if your degree is already underway. You may have to pay for an assessment, though, and it is worth asking your tutor or the disability support team at your university if there is any funding available. There is also plenty of advice on offer through support agencies, such as Dyslexia Action, which offer free advice sessions. For a fee, they can also carry out diagnostic assessments, screening and skill profiles, and arrange extra tuition.

WHAT DO STUDENTS SAY?

'I'm dyslexic, and so I was really worried about the amount of reading you need to do for a Geography degree. It nearly put me off applying, but then I realised that you have to do reading for any degree, so I decided to give it a go …. . Doing my degree has given me much more confidence and I don't think my dyslexia will prevent me from doing anything I want to in the future.'

(Donna, graduated with a BA (Hons) degree in Geography)

Summary points

1. Do not miss tutor meetings, and stay in close contact with your tutor throughout your degree. It is easier to discuss problems with your tutor if you both know each other.

2. If you are experiencing either academic or personal difficulties, do not wait for your next scheduled meeting with your tutor. Instead, email, text or phone your tutor to ask for an appointment.

3. Don't expect your tutor to know the answer to everything, and don't expect them to provide a 24/7 emergency response service! From time to time, your tutor may need to refer you to someone more specialised within the university who will be able to help you.

4. In order to make a smooth transition into university, it is crucial to arrange an early conversation with university staff about any learning support needs that you have.

CHAPTER 8

HOW WILL YOU BE ASSESSED?

Types of assessment

Let's start with the boring, but important, part. Your university is an educational institution with a legal charter entitling it to award its own degrees. In other words, in contrast to Geography at school or college, where external exam boards set the syllabus and the assessment papers, university Geography departments can design their own syllabuses, teach them and set and mark their own assessments. This does not mean that Geography departments are completely unconstrained in how they design their courses, as the *Subject Benchmark Statement for Geography* provides a set of criteria that all Geography degree programmes must meet. However, there is scope for variation between universities: for example, every university awarding a degree in Geography will have its own rules regarding the style and type of assessments and the marking criteria it uses to grade its students' performance. There will also be variation in the subject content of various degree programmes (see Chapter 2).

That said, broadly speaking, every module that you undertake as part of your Geography degree will assess your performance in two ways (see Chapter 3 for more guidance on modules):

Formative assessments: These are primarily to give you feedback on the quality of your answers, and do not count towards your final degree.

Summative assessments: These count directly towards your final degree. Therefore, as well as providing feedback on your summative assessments, lecturers will also provide a numerical mark.

Formative assessment can take a variety of forms. At one end of the spectrum, they can be very informal, such as a lecturer saying to a student who answers a question in their class, 'Good answer, well done'. It can be in the form of feedback from a lecturer on a draft of an essay that you are writing. At the other end of the spectrum, formative assessments might appear quite formal, such as a mock exam designed to give you practice before a real exam. For many students, the range of formative assessment used in universities will be similar to that experienced at school or college – think of all the sorts of tasks you were set by your Geography teachers as homework. Your homework was a type of formative assessment.

The range of summative assessment used by university Geography departments tends to be much broader than you will have experienced at school or college. At school and college, you typically encountered only three types in Geography: unseen exams; a geographical investigation; and maybe an EPQ (Extended Project Qualification). By contrast, a non-exhaustive list of the types of summative assessment that you might encounter as part of your Geography degree includes: seen exams; unseen exams; take-home exams; essays; assessed field and lab notebooks; dissertations; presentations; posters; reports; peer assessment (where the rest of the class have a say in your mark for an assessment); assessed seminars (where the quality of the discussion points you make is assessed); drawing GIS maps; or undertaking quantitative analysis. These are common types of summative assessment used by universities (although it is worth noting that potentially everything from this list could be set as formative assessment, as well). Some degree programmes will offer more unusual summative assessments on some of their modules, such as group essay writing, writing a letter to a newspaper, applying for a job in a geography-related field or making a podcast or website. So, as you can see, the various types of summative assessment that you could encounter are almost limitless, and many lecturers are very creative in the assessments that they set.

TOP TIP

Before you notify the Universities and Colleges Admissions Service (UCAS) which degree programmes you are interested in studying, make sure that you take into account the types of assessment used on your preferred degree courses. For example, some Geography degree programmes include far more unseen exams than others.

Understanding the marking process

In some assessments for university Geography (such as multiple-choice exams), the formula for working out the overall mark is easy. Each question is marked as right or wrong, and the marker adds up the correct answers to produce a final mark. However, the marking of more discursive assessments (like essays, essay-based exams, presentations, posters and reports) usually takes a more holistic approach. In other words, the final mark depends on the overall judgement that a marker arrives at regarding the quality of an assignment. Therefore, several assignments in a class could achieve the same mark, yet the marker could have reached it by weighing up a very different combination of strengths and weak-nesses prior to concluding that the overall impression is that they are both worth the same.

TOP TIP

Read the mark scheme thoroughly before each assignment and try to pick up clues from your lecturer about what they expect to see in your answers. The best way to do this is to speak to them one-to-one before you do the assignment and to ask them for feedback after the assignment has been marked (see Chapter 22 for further discussion of feedback).

This does not mean that the marking is completely subjective and at the whim of individual lecturers, as all degree programmes will have a marking scheme (some-times called grade descriptor, rubric, criteria sheet, grading scheme, scoring guide or similar). Put simply, these set down in writing what is expected from students to achieve a particular mark in a particular type of assessment (often, they are divided into 10 per cent increments). Although they vary slightly from department to department, marking schemes are usually quite detailed and show what is expected for each type of assessment (essays, presentations, exams and so on). You should look at the marking scheme for your department before you submit your first assignment so that you know exactly how your work will be marked. Table 8.1 can be used as a very rough guide to most types of assessment.

Table 8.1 Typical expectations in Geography assessments.

Marks of <40%	40% is the pass mark, at most universities. If your average mark for summative assessments is consistently below 40%, you do not understand most of the lecture material and should not expect to pass. This is quite rare.
Marks between 40 and 49%	If your average mark for summative assessments is consistently above 40%, you clearly understand most of the lecture material and should expect to pass.
Marks between 50 and 59%	If your average mark for summative assessments is consistently above 50%, you clearly understand all the lecture material and can recall it accurately. You should expect a 2.2.
Marks between 60 and 69%	If your average mark for summative assessments is consistently above 60%, you clearly understand all the lecture material and can recall it accurately. In addition, you have read some of the texts in the reading lists, understood this material and incorporated ideas from this literature into your assignment answers. You should expect a 2.1. Note that this indicates the importance of reading beyond lecture material, which is often fairly introductory in its level and tone (see Chapters 5 and 11).
Marks of >70%	If your average mark for summative assessments is consistently above 70%, you clearly understand all the lecture material and can recall it accurately. In addition, you have read widely beyond the lecture material, understood these texts and incorporated ideas from them into your assignment answers – and with depth, detail and originality beyond that expected for a mark in the 60s.

Undoubtedly, this approach to marking does rely on 'academic judgement', which means that various markers, in reality, may prioritise some of these criteria, in certain assignments. For example, some markers might expect perfect spelling and grammar in an assignment before they award the highest marks; others might accept some mistakes. Some markers might interpret differently the need to use a 'broad range of sources' to achieve the highest marks. Does it mean referencing five texts? Or ten? Or 15? While this may sound subjective, it is truer to say that the markers' interpretations of a departmental marking scheme for an assignment reflect what they think is reasonable, given its topic, type and length, and the time that students have to complete it.

That said, checks and balances will be in place to ensure that this approach to marking is fair and accurate. Some universities employ a moderation system, where a sample of marked assignments is double-checked by another academic in the same department (the moderator) to ensure that the approach that the marker has taken is both reasonable and consistent. Some universities ask for all assignments to be double-marked by two academics from the same department. Assignments are often marked anonymously, so that the marker does not know which student's assignment they are marking. Moreover, all British Geography degrees have external examiners (Geography lecturers from other universities), who scrutinise assessments and their marking for consistency and fairness.

WHAT DO STUDENTS SAY?

'I never bothered to read the mark scheme before my first essay and was devastated when I got 55%. I tried so hard and was expecting a mark in the 80s or 90s. I phoned my mum up in tears, as the written feedback basically said that I'd covered all the main points but had just summarised them from lectures, and that I needed to read more. Looking back, I realise that 55 was a decent start and that a mark of 90 in first year was so unrealistic.'

(Hannah, graduated with a BA (Hons) degree in Geography)

TOP TIP

Find out how your Geography department marks assignments and calculates degree classifications, so you know from the first year where you stand. For example, most universities do now allow you to re-sit an assignment that you have passed to try to gain a higher mark. Most universities do not allow you to request that an assignment is re-marked – you cannot appeal against what is often called an 'academic judgement'.

Finally, let us note that this approach to marking is very different from how essays are marked in school or college, where there is often a set list of points you need to make to answer the question, and you receive a mark for each one that you write about. This approach is good for consistency in an exam system where the teacher, the question setter and the marker are three different people. However, it tends to produce a lot of similar-looking essays, in other words essays that mention the key points to achieve the best mark possible. This approach does not reward the creativity or critical thinking skills that the lecturers who set and mark their own assessments can identify and that are key to achieving the best marks in university-level Geography (see also Chapter 17).

Understanding degree classifications

Ultimately, it is your degree classification that matters most. This is what is reported on your degree certificate and what employers will ask for. Most British universities follow the same classification system for honours degrees:

- First-class (sometimes called a 'first')
- Upper-second class (sometimes written as a 2.1)
- Lower-second class (sometimes written as a 2.2)
- Third-class (sometimes called a 'third')

Universities have various ways of calculating degree classifications from some or all of your summative assignment marks. Some will factor into their decision-making any personal extenuating circumstances that have impacted on your studies (see Chapter 7); others do not differentiate between the higher and lower levels of second-class degrees; a small handful also award 'double-firsts' for exceptional performance. It is definitely worth checking out how your Geography department calculates its degree classifications, so you know from the first year where you stand.

TOP TIP

Many students become obsessed with achieving at least a 2.1 (upper-second class) degree, as they believe that, in a competitive job market, you need at least this to get a graduate-level job. However, employers take into account other personal qualities and relevant work experience. Your degree classification alone is not a guarantee of success or failure in the jobs market.

Summary points

1. Whichever Geography degree course you undertake, you can expect to encounter a broader range of assessment types than you did at school or university.
2. The approach to marking is different at university. Consequently, you will need to change the way that you answer assignment questions.
3. Every university Geography programme can create its own marking criteria and its own rules for awarding degree classifications. Make sure that you understand these early in the first year so that you know what is expected of you.

Key reading

Weyers, J. & McMillan, K. (2011) *How to Succeed in Exams and Assessments*. Harlow: Pearson Education.

CHAPTER 9

WHAT TO EXPECT FROM THE FIRST COUPLE OF WEEKS AT UNIVERSITY

Making friends

If you move into a hall of residence (aka 'hall'), the first thing you will notice is the bedlam that surrounds you: hundreds of new students, all with luggage. Their journeys to university will have been more diverse than perhaps you would imagine. Some will have arrived alone from other parts of the United Kingdom; some will have travelled alone from overseas; others will be saying goodbye to parents, friends and large extended families while trying to get organised. Once the families depart, the process of trying to make friends can begin. For some, this process can be particularly unsettling, as a large group of strangers try to figure out who they will and will not get along with for the next few years. Do not expect to like everyone in your building and do not expect everyone to like you – that is not realistic. All you are looking for is a handful of like-minded people whom you can begin to call friends. This will involve some trial and error and, difficult though it is, the more people you meet, the sooner you will make friends. Rather than speed-dating, think of the first two or three weeks as speed-befriending, where positivity, personality and perseverance are the key. Do not worry too much about your new friends' backgrounds, as going to university is a great opportunity to meet people you would not normally get the opportunity to mix with. Also, do not worry whether or not your new friends are also studying Geography – most likely they will not be – as most students have a mixture of Geography and non-Geography friends.

TOP TIP

Feeling unsettled during your first few days at university is normal. Try not to spend too much time alone in your room. Instead, remember that everyone is feeling nervous and finding their feet, just like you! Get out and introduce yourself, no matter how daunting.

Whether or not you are living in a hall of residence or at home, some excellent opportunities to make friends will present themselves in the first couple of weeks – make sure you take them! An example is the inevitable Geography induction talk, given by a senior member of staff, followed by a chance to get to know fellow students over coffee or drinks. Many departments will run a Geography induction fieldtrip. Going on a fieldtrip with other students early in your degree increases the chances of making friends, as you are with the same group of people from your course for a longer period of time than is usual. Also, look out for the first meeting of each of your seminar groups (see Chapter 6) and your tutor groups (see Chapter 7), and make sure you attend. Both contain a relatively small number of students compared to lectures, and the chances of interaction and getting to know students in these groups increase.

Registration

You will arrive at university assuming that you are a student – but you aren't, not until you have filled in the registration forms and the university has processed these and added you to its systems. Some universities call this matriculation. What it means in reality is that, as well as making friends, you need to prioritise registering for your course. If you do not, you will end up stuck in a sort of limbo where you cannot swipe your student card to enter some buildings, you cannot access the university computers to check your emails or online timetable, you cannot access the sports facilities and so on. This normally only lasts a day or two while your registration is processed by university administrators, whom you will probably never meet, but it is still a disconcerting welcome to your new course.

WHAT DO STUDENTS SAY?

'I forgot to register for my course until the Friday of Freshers' Week. That was really bad, because the following Monday classes started but my form hadn't been processed, so my online timetable was blank, I wasn't on any of the lecturers' registers, and I had to miss the first session of the GIS module because I couldn't log onto the computers.'

(Dhruv, graduated with a BA (Hons) degree in Geography)

While the majority of modules are compulsory on most first-year Geography courses, some courses will offer you a choice of modules in the first year. These are sometimes called 'electives'. As part of the registration process, you will need to choose these – go to the 'module choice fair' to find out more information about the modules on offer before you make your decision. Also, make sure that the number of credits on the modules that you have chosen add up to the number that you are required to do in the first year of your course, as many universities will not accept your registration form if they do not. This would hold up your registration process even further! (See Chapter 3 for more guidance on modules and credits.)

Freshers' Week

A fresher is another name for a first-year university student and each students' union will make arrangements to welcome its freshers. Although these vary from university to university, often Freshers' Week takes the form of a week, before formal teaching starts, that is filled with a mix of social events, fairs and opportunities to get to know the campus and the city and to make new friends. As part of Freshers' Week, many universities organise a Freshers' Ball, which, despite its name, is generally an informal occasion involving an all-night DJ and/or a relatively famous live band. The ball is usually not restricted to freshers, so it offers the chance to meet students from other years.

WHAT DO STUDENTS SAY?

'Looking back, getting insanely drunk every night wasn't the best way to do Freshers'. Often, I couldn't remember who I'd been talking to the night before and was too hungover to take part in activities the next day .... I'd say you only have Freshers' Week once, so do loads of stuff, meet lots of new people and have a good time – but make sure you know your limits and don't drink till you put your health and safety at risk.'

(Ian, graduated with a BA (Hons) degree in Geography)

Another event is the Freshers' Fair, the key purpose of which is to enable new students to see what is available in terms of university clubs and societies. Clubs and societies use the fair to attract new members. The range is usually huge, so there will be something for everyone. However, as joining a society involves a fee, think carefully about what you would like to do and, realistically, about how much free time you will have. That said, if there is one, make sure you sign up to your Geographical Society, GeogSoc or similar – a society run by your students' union for Geography students. Whether you are living in hall or at home, make sure that you attend its induction events as they are excellent opportunities to meet other new Geography students informally. Although it can be hard at first if you are living at home, try not to make the distance between home and campus an obstacle to attending Geography social events, particularly in the first few weeks.

TOP TIP

A lot of university sports clubs hold their trials in Freshers' Week. If you want to be picked for a team, don't have a big night out beforehand!

Some universities include a part-time jobs event as part of their Freshers' Fair, while others run this separately. While most students have a part-time job while studying for their degree, again think about what is realistic before applying for any jobs. Finally, if you have moved away from home to attend university, while you are in signing-up mode it is also a good idea to register in Freshers' Week for a new GP. This means that your university town is where your permanent doctor is located and, when you go home, if you need something, you just register with your doctor there as a temporary patient.

The main activities of Freshers' Week will probably have been organised and run by your students' union. Your local students' union is at the heart of the social life of the university, and you can obtain a union card once you have registered on your course. Students' unions are run by students, for students. Apart from organising Freshers' Week, they offer students advice, support and representation, although the exact nature of this varies according to the size of the university. If you want to become more actively involved in your students' union, you could consider becoming a union representative, contributing to the university newspaper, magazine or radio or helping with Rag Week.

Lectures, seminars and tutorials

The week after Freshers' Week, teaching starts properly. Many universities have an online timetable or an app where you can look up your timetable. We have already talked a lot in Chapters 5 and 6 about what goes on in lectures and seminars so, here, let's just note that your very first lecture will feel strange!

You will find yourself sitting in a huge room, surrounded by people you do not know, listening to a lecturer talking about things you have never heard of. But do not worry, it is all part of the experience and soon you will feel right at home. The other thing you will do in the first week is to meet your tutor. Hopefully you will remember from Chapter 7 that this is a person whom you will meet regularly for one-to-one conversations. They will be your go-to person for support and guidance, and they will write references for you – so, strive to make a good first impression. We are both tutors, and we are astonished when students turn up for their first meeting with us still hung over or, worse still, looking a bit dishevelled and on their way home from a night out!

You will be lucky if much of your first-year teaching takes place in the Geography department itself. To make best use of the teaching rooms, most universities timetable the classes across the whole campus. Whether you are living in hall or attending the university in your hometown, you should expect to get lost on campus – a lot. Some university campuses are tightly packed, with a large number of buildings in a relatively small space, while others have buildings spread all over the town or city centre. Potentially, this means plenty of walking between classes and opportunities to end up late at your first class! Your lecturers will expect this in the first week, but not in the second.

> ## WHAT DO STUDENTS SAY?
>
> 'A lot of students party hard for the first two or three weeks, but you've then got to tell yourself when it's time to slow down and rebalance – no one else will tell you!'
>
> (Becca, graduated with a BSc (Hons) degree in Geography)

Money worries

For most Geography students, the first week of university will be the first time that they have been in control of their own finances. Living on a budget means living within your means and keeping track of how much money you are spending every week. Most students find that in their first week at university they spend more than they had planned to. That is okay, as long you are spending sensibly from Week 2 onwards. Work out how much money you have available for your first year and give yourself a limit of how much you can spend each week. Ask other students about the best places to shop, so you are not constantly buying the most expensive bread and milk in town just because the corner shop is handy!

> ## TOP TIP
>
> Don't miss the first lecture of each module, as this explains what the module is about, when and where teaching on the module takes place, what assignments to expect and so on. Missing the first lecture will leave you clueless about what to expect in the weeks to come!

For most Geography students, a student loan is essential. It is better if you have sorted this before university starts, but you can apply once teaching is underway (see the Student Loans Company's website). However, to receive your first instalment, you need to register for your course – which is another reason to fill in your registration forms as soon as you arrive! Many Geography students have a part-time job. The money is helpful, and it is good for their curriculum vitae (CV). Jobs with some flexibility in the required working hours are best. Jobs like working in a club, which involve regular late nights, are best avoided.

> ## TOP TIP
>
> As your student loan won't come in straightaway and there can be delays, don't arrive at university with no money in your pocket!

The urge to visit home or quit your course!

Some first-year Geography students immediately love their new university, course and lifestyle. For others, it takes more getting used to. If you have moved away from home to attend university, you may become homesick. If you do not feel that you can confide in your new friends, your university will have counsellors – look on the university website under 'welfare'. You should have been assigned a personal tutor, who will be able to offer support and advice (please read Chapter 7 for further discussion on the role of your tutor).

If, after attending some of the classes on your course, you decide that Geography is not for you, it is best to let your tutor know as soon as possible. First, give this careful consideration. At most universities, you can change your degree course during the first couple of weeks of the academic year, but you do not want to change and then realise that Geography was the right course after all! There will be a lot going on in the first few weeks – new town or city, new friends, new university, new course, new independence, new routines – and you need to give yourself a chance to adjust and for things to settle down before you make any hasty decisions. If you are still getting used to some of the other changes, changing your course will probably not be the solution that you hope it will be.

WHAT DO STUDENTS SAY?

'Even if you are feeling homesick, try to avoid going home for the first three to four weeks of the first semester. Skype or FaceTime your friends and family instead. The more time that you spend on campus and in halls, the more likely it is that you will make friends and get used to your new surroundings.'

(Claire, graduated with a BSc (Hons) degree in Geography)

Summary points

1. Do not panic if the first couple of weeks seem overwhelming. It will soon all become familiar.
2. Enjoy Freshers' Week and make the most of this opportunity to make friends. Remember, at first everyone is feeling nervous and finding their feet, just like you!
3. Do not miss your first meeting with your tutor (it creates a terrible first impression) and do not miss the first lecture of each module, as this explains what the module is about, when and where teaching on the module takes place, what assignments to expect and so on. Missing the first lecture can leave you confused about what to expect in the weeks to come!

PART II

THE ACADEMIC SKILLS YOU NEED TO SUCCEED

CHAPTER 10

LISTENING SKILLS

Why is listening effectively important?

How often do you think about your listening skills? We would bet quite a lot of money on your answer being 'Not very often'! The reason we are so confident is that there is a tendency for new students to focus on improving their reading and writing skills first, so that they feel confident with their early university-style assignments (see Chapter 8). Next, they typically focus on developing their public- speaking skills, as the attention in seminars and tutorials (see Chapters 6 and 7) tends to be on students who talk the most and appear to be the most confident in contributing to group discussions. Here, it is easy to fall into the trap of assuming that verbally confident students are the most engaged students.

In reality, students who are skilled listeners are likely to learn better than those who prefer to speak a lot of the time without giving themselves time for reflection. So, in this book we have included a chapter on listening, as we know that good listening skills are something that can help you to succeed at university. We feel that more attention should be given to this aspect of learning. First, we discuss the various types of listening; next, we discuss the barriers to listening that you need to be aware of, so you can work to overcome them; and, finally, we discuss the importance of what is known as 'whole-person listening'.

WHAT DO STUDENTS SAY?

'I don't think I was very good at listening when I started university. I was always asking questions that apparently my tutor had already told the class! I worked on this a bit and found that to really listen properly – and to think about what was said in lectures and in readings at the same time – was hard work and took a lot of effort. I found it quite tiring when I had quite a few classes on the same day, so started to have an early night before those days so that I was prepared for it.'

(Molly, graduated with a BA (Hons) degree in Geography and Planning)

Types of listening

When it comes to listening, people often talk about 'active listening'. This is about listening so that you hear *what* is being said, while also considering *how* the words are being spoken. This means considering things like tone, volume, expression and so on. Active listening is also about observing non-verbal aspects of communication, such as body language and other behavioural cues, which help you to interpret the meaning of what is being said. It is useful to be aware of active listening and what it is, but there are many other types of listening as well.

Discriminative listening (identifying)

This is the most basic type of listening, where you identify sounds and discriminate between them. Discriminative listening can also be about detecting emotional variations in a person's voice, their various emphases in intonation and how their body language helps you to identify what is being said.

Comprehensive listening (understanding)

This moves beyond listening that is about identifying issues. It is about listening to understand, and it is also sometimes called 'content listening', 'full listening' or 'informative listening'. All the skills involved in discriminative listening are needed here, as is an understanding of words and grammar so that you can have a full appreciation and understanding of what is being communicated.

Analytical listening (reasoning)

This is about listening in order to critically evaluate what is being said and to form a decision or reach a conclusion about the specific issue at hand. It is sometimes called 'critical listening', 'judgemental listening' or 'evaluative listening'. This type of listening can be very hard work! It requires a lot of concentration and effort to listen carefully and evaluate what is being said (and what is not being said) in the context of existing knowledge and understanding. If you can develop this type of listening skill in the first year of your degree, the chances are you will get a lot more out of your lectures and seminars.

Empathetic listening (feeling)

This is about listening both to understand and appreciate how a person feels about the issue or topic that they are talking about. High-quality empathetic listening can result in you feeling what that person is feeling (whether this be joy, delight, anger, frustration or confusion). People are more likely to talk about how they feel if we use open body language gestures that help them to feel confident in sharing such information. This is something called sympathetic listening.

Appreciative listening (liking)

This is about listening and enjoying or appreciating the message of the speaker. Appreciative listening tends to be used when we hear an inspiring leader or moti-vational presentation, as well as when we play our favourite music.

Barriers to listening

In addition to the types of listening listed above, there are other types of listening that it can be useful to be aware of, as they can prevent or restrict effective learning. What is sometimes referred to as 'false listening' is when we give the impression that we are listening yet we are not actually hearing anything that is being said. 'Preliminary listening' or 'initial listening' is when we really listen only to the points made at the start of a lecture or presentation; there then comes a point where we 'switch off'. This is similar to what is sometimes called 'partial listening'.

Finally, 'selective listening' or 'biased listening' is where we listen only for specific points or issues – normally those that we agree with – and ignore anything else that is said. For example, you may miss important information in lectures because you find yourself subconsciously being more likely to listen closely to older, male lecturers than younger, female lecturers. Likewise, throughout your Geography degree you will encounter topics that you have already made your mind up about. At these times, adopting a closed mindset – either because you think you know the topic already or because you have taken up a specific position on the topic – can result in problematic 'selective listening'.

WHAT DO STUDENTS SAY?

'I was quite surprised by some of the issues that we were taught in Human Geography in first year. Some of the topics were very political and, to be honest, quite controversial. I realised I didn't really know a lot of the background to many of them and the different sorts of explanations for why things are the way they are, if you see what I mean. I think what I'm saying is that I was a bit biased in my points of view. I think it was useful for me to recognise this so I could pay more attention to what was being said and listen properly to different viewpoints .... Yeah, I think I've changed in some ways and definitely have different views on some things now.'

(Stephen, graduated with a BA (Hons) degree in Geography)

Another significant barrier to listening effectively is interruption. Interrupting someone who is speaking – say in a seminar or tutorial – implies that you know better than them and that your words are better than theirs. Often, people who interrupt another are incorrect in their assumptions about what they will say. It also is not very nice to be interrupted, especially when you have a point to contribute to the discussion and know what you want to say.

Physical barriers to effective listening

- Competing noises: other conversations, phones ringing and so on.
- Environmental constraints: being too hot or too cold or feeling uncomfortable.

Mental barriers to effective listening

- Jumping to conclusions about the subject matter (resulting in not listening properly).
- Preferring to talk rather than to listen.
- Focusing on the delivery style of the lecturer (their manner, approach, dress and so on) rather than on what is being said.
- Preoccupation with something else (such as your phone, email, Instagram).
- Being disengaged from the subject.

Emotional barriers to effective listening

- Feeling strongly about the person who is speaking (e.g. dislike, fear).
- Feeling strongly about the subject matter (e.g. sadness, anger).
- Being preoccupied with emotions about something else (e.g. last night's hangover!).

Figure 10.1 Some barriers to effective listening.

TOP TIP

If you find that your listening skills are not as effective as you would like them to be, consider if there are any physical, mental or emotional challenges or obstacles that you could address to help you to improve.

Whole-person listening

As discussed in Chapter 11, some students focus on writing down everything in lectures, maintaining a neat and tidy set of notes that combines everything the lecturer has on their PowerPoint slides with everything that they have said during the lecture. Essentially, this is about copying down text, so it is not a very sophisticated approach to listening or learning. There is little time to listen and think about what the lecturer is saying if your focus is on trying to produce a transcript of the lecture.

A much more sophisticated form of listening is 'whole-person listening'. This is sometimes called 'deep listening'. It takes active listening to a new level, where factors such as emotion, body language, identification of biases, beliefs, values, practices and other issues are all considered in the process of listening. This approach is essentially about understanding the complete person behind the words, and so is about seeking to fully understand your lecturer or seminar leader, beyond the words that they say, and to appreciate their emotions, values, beliefs and attitudes through a close reading of their words alongside their body language, emotion, pace and demeanour. It is one thing to be careful about listening and to record what is being said; if you listen more deeply, your listening could move up to a new level.

TOP TIP

Focus on deep listening to analyse the meaning of what is being said rather than listening only to identify and understand issues.

Summary points

1. As well as active and deep listening, you can employ *discriminative listening* (to identify), *comprehensive listening* (to understand), *analytical listening* (to reason), *empathetic listening* (to feel) *and appreciative listening* (to like).
2. There are a number of barriers to effective listening, including challenges that are physical, environmental and emotional.
3. Focus on appreciating the deeper meaning and message being communicated.

Key reading

Kline, N. (1999) *Time to Think: Listening to Ignite the Human Mind.* London: Octopus Books.

CHAPTER 11

EFFECTIVE NOTE-TAKING

Why do students take notes?

An important skill to develop in order to succeed in your Geography degree is effective note-taking. This is particularly important when it comes to getting the most out of lectures (see Chapter 5) and also when it comes to undertaking fieldwork (see Chapters 20 and 21) and becoming an effective academic reader (see Chapter 13).

A key question to ask yourself is why you are taking notes. You may have various reasons: to help you to get a grasp or overview of the topic or issue; to help you to choose material that is relevant; or to enable you to recall information. Taking notes is important, as we are all more likely to remember and understand key points if we have written them down rather than are relying on simply hearing them or reading them. Taking notes can also help to ensure that you are actively listening to your lecturer or are engaging with your reading, not just passively letting things drift by you. Your notes also act as a useful point of reference after you have finished the session. You can take notes for any one of these reasons, or a combination of two or all of them.

A common mistake that many new Geography students make is to assume that they need to write everything down, almost as a word-for-word transcript of each class they attend. However, more often than not you will not have time to do this while listening and thinking about what is being said and making connections between ideas and things that you have read. In reality, the points discussed in a lecture or seminar will also vary in their relative importance, so you will not need to write everything down.

Another point to consider is whether to use the traditional approach of pen and paper or instead to use your laptop to take notes. This depends very much on your personal preference. Many university lecturers make their PowerPoint slides available to students before each session. Students who type their lecture notes during a lecture are usually editing this copy directly, to add in extra information as it is spoken by the lecturer, rather than keying in *all* the information. For this to work well, you still need to be relatively good at typing!

> **TOP TIP**
>
> Ask yourself why you are taking notes and shape your note-taking technique to fit your motivation.

Note-taking techniques

Ultimately, in the first year of your Geography degree you need to experiment a little to find a note-taking technique that works well for you. However, as a starting point, it is worth considering some of the standard techniques that you can adopt when it comes to note-taking – you may even want to consider adopting a separate technique for the various types of note-taking. For example, adopting the approach of thematic bullet points might work well in lectures where the lecturer is structured and systematic, but, when note-taking for an essay, you may use verbatim sentences or mind-mapping. It is important to develop note-taking techniques that work for you, and it may be that you adopt very different techniques from those used by your friends or the others on your course.

Alongside these varied approaches, some specific techniques can be useful when it comes to effective note-taking. Consider using **bold** or BLOCK CAPITALS or colour or underlining to emphasise key points. Use your own words rather than simply repeating what the lecturer or seminar leader says, as this will help you to think more deeply about what is being said. Use abbreviations to help you to keep pace with the class. Be careful to identify the main arguments and sub-points clearly. If handouts are provided, annotate them and integrate them into your approach to note-taking. Make sure that you make a note of the author, year and page number of any direct quotations (see Chapter 16 for more guidance on referencing). Lastly, keep your notes from your reading filed with your class notes, as this will help you to make connections between the two (see Chapters 4 and 8 on the importance of not relying just on your notes from class).

Verbatim sentences

This approach is simply about recording key points, either in sentence form or in bullet points, with each point being given a new line. This approach is useful for recording key points, particularly when the material is delivered quickly. There are likely to be key words that are used regularly, so having a shorthand form of these will help you to keep pace with the class.

Cornell note-taking system

When using this approach, you divide each A4 page into three sections: a column roughly 5 cm wide on the left of each page, to be filled in after the class; the wider column of about 15 cm remaining on the right side, where you take your main notes during the class; and a summary section on the bottom quarter of each page. The left-hand column, completed after the class, is where you note your key questions or terminology, or the connections to things that you have read or things you have learnt in other classes. This approach can be useful in understanding key ideas and associations.

Mind-mapping

Using a mapping approach to note-taking is visual, and it can take many forms. One of the most common approaches is the spider diagram, where you map out key themes from a central topic and add additional sub-themes and issues as the class develops. Other approaches here could include a timeline approach that maps changes across time, or an argument-based approach that maps out key sets of arguments about a specific issue. This approach is useful for visualising ideas and recognising the connections between issues. This can also help you to hold together a lot of separate information and can help to focus your attention, given the connections between ideas.

Thematic bullet points

This approach is characterised by key themes or headings, with bullet points under them to provide more detail and content. So, for example, you might have the main topic heading and then a set of key subheadings for the themes related to each one. Within each sub-theme, you could have several bullet points to provide additional detail and information. You can indent the sub-themes slightly as a way of streamlining your notes. This approach is useful in developing a systematic set of themes in an organised way.

Figure 11.1 Common ways to structure your notes.

WHAT DO STUDENTS SAY?

'I used to take my lecture notes and then file them after the class, and not really look at them again until we had an assignment due or an exam. It took me a while to learn that it was better for me to look at my notes as soon as possible after the lecture, either to add to them with points from my wider reading or to identify quickly things I didn't fully understand.'

(Suzanne, graduated with a BSc (Hons) degree in Geography)

Organising your notes

As we noted in Chapter 5 on 'Getting the most out of lectures', it is important to note the date, the name of the module and the name of the lecturer at the start of each class, as this will help you when it comes to organising your notes, integrating them with your other notes from the same module and, if you need to clarify any points at assessment time, asking the appropriate member of staff. Remember that when it comes to note-taking, you need to develop a system that means that your notes are useful to you weeks or months after a lecture or seminar and are immediately understandable. This is where it can be helpful to leave some space in your notes to add additional points that you may have learnt about as your modules progressed or that you realise you may have missed at the time.

TOP TIP

Be organised by deciding on a filing system for your notes that works for you and stick to it (e.g. one lever-arch file per module, one notebook per module or one folder on your computer per module if you decide to take all of your notes electronically).

Summary points

1. Effective note-taking is an important skill to develop; you are more likely to be able to recall material if you have taken notes about it.
2. Your notes will be more effective if you know why you are taking notes and alter your approach accordingly.
3. There are a number of approaches to note-taking, and you should adopt the approach that is best suited to your style of learning and your reason for taking notes (e.g. lecture, assignment).
4. Consider the various techniques for making your notes clearer, such as type-face, colour, highlighting, underlining and so on.
5. Integrate the notes from your wider reading and any handouts from classes into your note-taking approach, so that you can see the connections between your notes from various sources and any additional information provided in the class.

Key reading

Waller, R. & and Schultz, D. (2015) *How to Succeed at University in GEES Disciplines: Enhancing Students' Information Literacy Skills*. York: Higher Education Academy. Available at https://www.geography.org.uk/download/ga_conf15_forumhandout2.pdf

CHAPTER 12

APPROACHING READING LISTS AND LIBRARY SEARCH STRATEGIES

Getting started

You might be thinking that this chapter is not necessary, as we should all know how libraries work. Before you skip a few pages, however, stop and consider whether university libraries could be different from the public libraries or school libraries with which you are familiar. The short answer is yes, they are different. This means that we do really need this chapter!

Let's start with some basics. Definitely familiarise yourself with your particular university's library in the first week or two. Make sure you know how to use the online library catalogue to search for texts. Check you know where the Geography books are shelved in the building and how to take books out of the library. These are important first steps, before you have any assignments. Too many students leave it too long to master these basics. Helpfully, most (but not all) libraries use the same numbering system to organise their books. Geography books will be on the shelves labelled with the numbers 910–919. However, because Geography overlaps with some other subjects, shelves starting with the numbers 300–399 will contain books that are useful to some Human Geography topics. Some useful books for Physical Geography will be on the shelves labelled 550–560. Make sure that you know where these shelves are located in your library.

WHAT DO STUDENTS SAY?

'With hindsight, I wish I'd gone on a library induction tour in Freshers' Week. It sounded really boring, but it would have saved loads of time in the long run, as the tours didn't run after the first couple of weeks and lecturers just assumed you'd know how to find stuff.'

(Gemma, graduated with a BSc (Hons) degree in Physical Geography)

Apart from books, in your university's library you will also find journals (aka periodicals), reference materials, newspapers, magazines (like the *Economist*, the *Spectator*, etc.), DVDs, maps, quiet or silent study areas, group study areas, computers, printers and photocopiers (to photocopy any texts that you cannot take out of the library). It is books and journal articles that you will mainly use in your Geography assignments, as most lecturers will want you to engage with sources that have been written by academics and then peer-reviewed (in other words, checked) by other academics before publication. If you are not sure what a journal article is, do not worry – most new students will not know either, so we will talk more about this a little later in the chapter. For now, we will just make a note that academic books and journal articles are important, and that we should not be fooled into thinking that popular magazines, newspapers and websites are good alternative sources of reading for university assignments – they are not. However, depending on the topic that you are studying, there is also a whole range of other types of outputs that you could also engage with from time to time, such as policy reports from government departments, think-tanks, non-government organisations (NGOs) and charities.

Reading lists and module guides

We can both remember the horror we felt when we encountered our first reading lists at university. For both of us, it was during the second week of university. Freshers' Week had passed, and 'being a Geography student' was starting to feel very real. Some lecturers gave us reading lists for the entire year as part of their 'module guide' (a kind of instructional manual for how they would teach their module) at their first lecture; some gave us A4 sheets of paper each week, with seemingly endless lists of sources; and others put their reading lists online.

The thing we know now, which we did not know then, is that Geography lecturers do not expect their first-year students (or indeed any students) to read everything on these reading lists. However, no one told us. We did not manage it and, if you try, you will not manage it either. Just as our lecturers did, we, too, now feel compelled to set long reading lists, as our class sizes are large and the number of copies of each text in the university library is limited, therefore a long reading list for each class hopefully ensures that everyone in the class can read at least some of the texts on it each week. As discussed in Chapter 4, at university more independent study is expected – but no student should try to read everything on their reading list.

WHAT DO STUDENTS SAY?

'I remember feeling completely overwhelmed in the first few weeks of first year and thinking "Bloody hell, how am I ever gonna read all this?" Then the penny slowly dropped as first year progressed. At school, there was mainly teaching and some homework. All that had changed at uni was the balance. So, to get good marks, there's more independent learning required. But that doesn't mean reading everything cover to cover.'

(Stephen, graduated with a BA (Hons) degree in Geography)

So, while you should not feel intimidated by the length of the reading lists with which you are confronted (though, to some extent, nearly every new student is), you should notice that we said that you should try to read 'some of the texts' (plural). You may well need to read more than one text on the reading list every week for every class in order to construct effective arguments and to think critically in assignments (see Chapter 17). That said, it is often unnecessary to read every source that you choose from start to finish (see Chapter 13). If this thought has you panicking, hang in there (at least until Chapters 13 and 17!). For everyone, this is all very unlike how you read about geography before you came to university. There is definitely a knack to reading geographical texts at university quickly and effectively, and you will master it with practice.

TOP TIP

Try to take books on a reading list out of the library before each lecture starts. After the lecture, the demand for books is always higher and so the chances of having to join a waiting list increase.

Books and journal articles

On your reading lists you will find mainly books and journal articles. As you are currently reading one, you might think that we do not need to spend time defining what a book is! However, there are three types of books that you will use at university: textbooks; monographs; and reference books. Textbooks will be most familiar to you. They are what you might have been given at school or college to study Geography. They tend to give a general introduction to a subject, and so are written in a (hopefully) accessible style and without getting into too much complicated detail. This book is a textbook.

In contrast, monographs are specialist books, normally written by an academic about a piece of research that they have conducted on a specific topic. They are aimed at an audience of undergraduate students, postgraduate students and other academics, so they assume some prior knowledge. Often the authors of monographs make a single, clear argument in their book or have a particular viewpoint on the subject – and you normally do not need to read the whole book to figure out what this is!

Reference books – such as the *Dictionary of Human Geography* (Gregory et al., 2009) – tend to be useful when trying to understand some of the more specialist vocabulary used in university Geography, but they do not offer enough depth to be used commonly.

All three types of book may be written by one or more authors (like this one). Alternatively, each chapter could be written by a different person. These are called edited books, where the job of the editor, or editors, is to make the various chapters fit together so that the book flows well when read from cover to cover.

The key problem that arises is that (unlike school or college textbooks, which tend to have been written specifically for the syllabus of your exam board) none of these three types of books will have been written specifically for the degree course that you are now studying. The content of all Geography degree courses depends on the research interests of the staff who work in each department and the modules they run (see Chapter 2), to some extent.

The benefit of the course content depending on staff's research interests is that you get cutting-edge teaching. The downside is that it is not economical to publish a book for every module or degree course. Even if it were, it would not be desirable. If you can find two or three monographs on the same topic, you can compare, contrast, criticise and evaluate the various arguments and viewpoints of the authors for assignments (see Chapter 17). These are the skills that university markers reward (see Chapter 8). Textbooks and reference books tend not to make arguments, but instead present information in a simpler form, sometimes as a series of uncontested facts. That is why textbooks are sometimes recommended reading for first-year students as an introduction to the topics you will study as part of your degree. However, as your degree progresses, you will inevitably end up reading more and more monographs. The challenge is to pick your way through the sometimes difficult and detailed analysis that they contain to identify the key points that the author is trying to make (see Chapter 13).

TOP TIP

School or college has finished, so don't be tempted to use Wikipedia or your old, familiar textbook in an assignment for your degree. Instead, use books and journal articles from the university library.

Moving on, what is a journal article? Or, for that matter, what is a journal? Think of a journal/periodical as being like a magazine on a specialised topic, which is published on a regular basis several times each year. For example, *Hello* magazine is published weekly and focuses on celebrities (their marriages, babies, affairs and large houses). In contrast, the *Transactions of the Institute of British Geographers* is published four times each year and addresses current issues in Human Geography. While *Hello* clearly wins the contest for the best title, if you dig deeper, they have similarities. They both contain lots of fairly short articles, written by different people, about unrelated but contemporary topics of interest to their readers. They are both peer-reviewed to make sure that they are not inadvertently talking complete nonsense – in the case of *Hello*, the articles are reviewed by the editor, and in the case of *Transactions* they are reviewed by other academic geographers. And, due to its narrow focus, each article is fairly quick to read – though you can reasonably assume that a *Transactions* article will take longer than an article from *Hello*.

So, what's not to love about journal articles? They are shorter than books, more relevant and up to date, while also being highly credible by presenting the findings from original research. It is a condition of publication in a journal that the material for the article has not been published elsewhere. Adding to their appeal is the fact that nearly all academic journals are now online, so can be read anywhere on a computer, once you have been given access by your university.

WHAT DO STUDENTS SAY?

'Don't rush out and buy lots of textbooks in the first week of the year. Academic books can be really expensive, and you probably won't use the same books every week anyway. Instead, use the books in the library and try to get into the habit of reading journal articles and e-books online.'

(Tariq, graduated with a BA (Hons) degree in Geography)

Searching for texts which are not on your reading lists

At first, just sticking to some of the texts on your weekly reading lists will be fine. However, once assignments (particularly essays) have to be written, you need to head down to the library to find your own sources (academic books and journal articles) to supplement the texts on your reading lists. Why? Because if you only keep to the reading lists, you will be reading the same introductory sources as everyone else in your class. If you read the same introductory sources, you will end up making the same broad points as others in your class; and if you make the same broad points as others, your essay will lack the 'wow factor' needed to get the higher marks. Geography degrees tend to reward independent thinking in assignments and the creative use of sources to make an interesting argument – and this can only be achieved by reading a wider range of texts than are provided on reading lists (see Chapter 8 for a more detailed discussion of this).

TOP TIP

Reading lists are only your starting point. You also need to read more widely.

So, where do you begin? One starting point is to look at the bibliography – the list of sources (also known as the list of references) – at the end of each of the texts on your reading list (see Chapter 16 for guidance on referencing). You can find in your university library some of the items listed in these bibliographies; read them, and then repeat the process by looking at their bibliographies, and so on. Eventually, you will find that some authors' names keep recurring in all the bibliographies that you have looked at. This is a good thing, as it means that you are starting to identify the 'big names' who have written the most about this topic. You will be expected to discuss their work in your assignments.

Your next step is to access your university library's online catalogue. Each university's catalogue will have its own nuances that you need to get to grips with quickly. However, broadly speaking, they all allow you to search all of the books and journal articles in the university library by an author's surname, by title or by key words in the title. If the item that you want is held as an electronic copy, the best library systems take you to its full text with a single click. If it is not, the catalogue will give you the book or journal's shelf number where you can find the paper copy.

The huge advantage of these systems over Google and Wikipedia (or, indeed, a public library) is that you know that all the books and journal articles have been written and peer reviewed by academics and so will be useable in assignments. However, bear in mind that library catalogues are very sensitive to things like American spellings (for example, if you are doing a geopolitics essay, searching for 'defence' and 'defense' will give you two completely different sets of results). Most library catalogues help you to cut down or to increase the number of sources that each of your searches finds by allowing you to add the words *and*, *not* or *or* into your search terms (for example, searching for any texts with the words 'defence *and* geopolitics' in their title will give you completely different results from searching using 'defence *or* geopolitics'). If you are still struggling to find sources, get a thesaurus and try typing in synonyms, as most libraries' catalogues are capable of matching the words that you type only to the exact same words in the titles of the books and journal articles that they hold. Most catalogues are no more intuitive than that. No one said that this was going to be a quick process!

TOP TIP

Always check that the sources you are using have been written by academics. Also, check when they were written, as academic texts can go out of date in a few years. You need to be using the latest information in your assignments.

Summary points

1. If there is one on offer, go on an induction tour of your university library in Freshers' Week. Make sure you know where the shelves of Geography books are located.
2. Spend time becoming familiar with your university's online library catalogue, e-books and online journal articles. You should know how to access these before your first assignment is due.
3. Make sure that any text that you find that is not on any of your reading lists is written by a credible, academic author.
4. Get into the habit of going to the library regularly to study. It is difficult at first, but soon becomes second nature.

Key reading

Healey, M. & Healey, R. (2016) 'How to conduct a literature search.' In: Clifford, N., Cope, M., Gillespie, T. & French, S. (eds.) *Key Methods in Geography*. London: Sage.

Waller, R. & Schultz, D. (2015) *How to Succeed at University in GEES Disciplines: Enhancing Students' Information Literacy Skills*. York: Higher Education Academy. Available at https://www.geography.org.uk/download/ga_conf15_forumhandout2.pdf

Deane, M. (2010) *Inside Track to Academic Research, Writing & Referencing*. Harlow: Pearson Education.

Dictionary of Human Geography. Available at https://www.wiley.com/en-gb/The+Dictionary+of+Human+Geography%2C+5th+Edition-p-9781405132886

CHAPTER 13

BECOMING AN EFFECTIVE ACADEMIC READER

Reading skills

As discussed in Chapter 12, Geography lecturers do not expect their first-year students (indeed, any student) to read everything on the reading list that they give them. A long reading list for each class hopefully ensures that each week every student can read at least some of the texts on it. However, we also said that relying exclusively on reading lists is not always enough and that, when it comes to preparing for an assignment such as an essay, you need to seek out additional sources from the university library. In reality, this means that you have to read a wide range of sources in order to construct effective arguments and to think critically in assignments (see Chapter 17).

So, how do you go about reading academic sources? While the answer to this question may seem obvious, in reality there are many tips and tricks that you can use to save you a lot of time in the years to come.

As a starting point, it is useful to ask yourself why you are reading. It may be that you are reading for a seminar, a presentation, an essay, a group report or exam revision. Having in mind a clear set of reasons will help to keep you focused and on track and to identify the readings that are most relevant and appropriate for your purpose. As it is not possible to read everything, you need to work on being able to identify quickly what to read, what to partially read or fully read and what to avoid reading.

WHAT DO STUDENTS SAY?

'I was overwhelmed when one of my lecturers gave out a massive course guide to us! It was packed full of lists of readings and I didn't know where to start. I realised after a while that we weren't expected to read everything from cover to cover. Panic over!'

(Bunmi, graduated with a BA (Hons) degree in Geography)

Skimming and scanning

Two techniques that are useful to develop are skimming and scanning. Skimming is where you quickly survey the text that you are about to look at. You flick through the pages, skim the contents page, look at the chapter or section headings and note any specific points that jump out at you or any sections that you want to go back and read in more detail later. This is a useful way of getting a general feel for the text so that you can decide if this is something that you should read. It can also be useful to skim over the abstract and quickly survey the reference list or bibliography at the end to get an idea of the kind of author that the article or book you are reading engages with.

Scanning takes this a step further, and is about generating a sense of what the text you are engaging with is about, so you can assess how useful it may – or may not be – to the task at hand. When scanning, you might read the introduction and conclusion a bit more closely than you did when you skimmed through it. You could read the first and last sentences of various sections to get a slightly more detailed idea about what the text is about. If there is an abstract, you might consider reading this more closely to get a feel for the article, so that you can decide if it is worth reading all or parts of it more closely.

The final step here is to identify the key points from the text that you have been skimming and scanning. If it is a journal article, a close look at the abstract and a couple of other sections could help you to identify some of the key points that the author makes. Some questions to ask yourself here include: What is the argument of this article? Does the author use a specific theory or take a particular perspective on the issue? Is the article reporting on research and, if so, are there some key findings that you should be aware of?

TOP TIP

Unlike a novel, you will rarely read an academic book or journal article in its entirety from start to finish. With practice, you will develop the skill of identifying the key points quickly and without reading the whole text.

Survey, question, read, recall and review (SQ3R)

Another well-known technique for developing your academic reading is the 'survey, question, read, recall and review approach' (shortened to SQ3R, for convenience). Adopting this approach could help you to develop into an effective and quicker academic reader. Figure 13.1 shows how it works.

Survey: This is about skimming and scanning the text, as discussed above. It gives you a general feel for what the reading says so that you can assess how useful it will – or will not – be to your work.

Question: Ask yourself if this reading will assist you. Does it provide ideas or information that is useful for you and for your essay, seminar or project? How does the argument that it makes compare to other things you have read? Asking yourself this will help you decide whether to read the text in more detail.

Read: If you make the decision to read the text more closely, this step is about reading through only the relevant subsections of the text carefully, making notes of relevant points while remembering your main purpose in reading the text.

Recall: This is where you put the text and any notes that you have made to the side to see what you can recall about what you have read and written (see Chapter 11 on effective note-taking to help you here).

Review: Finally, review your understanding and perhaps re-read part of the text to check that you have fully understood the argument. This can help you to clarify any points that you were unclear about or may have misunderstood.

Figure 13.1 SQ3R approach to reading.

This technique can be useful to apply as you work to improve your critical academic reading skills. Many students tend to pay most attention to the first 'R' listed here – 'read' – without paying much attention to the others. However, to improve your reading skills and critical thinking (see Chapter 17), the points about surveying and questioning the text are crucial, as are recalling and reviewing what you have read.

> ## TOP TIP
>
> Remember SQ3R, apply it to the next five readings that you engage with at university and then consider how your reading skills have developed as a result.

Speed-reading

Speed-reading is a term used often to describe a set of techniques for reading through a text quickly, using some of the methods that we have already discussed. In addition, an approach that can be useful is finger tracing. You run your index finger across the page as you read the text. You can alter the speed at which you do this, to see what you notice as you move through the text. Alternatively, you can try continuously moving a blank strip of A4 paper down the page you read at a steady pace, without stopping, revealing each line of text in turn. This will force you to focus on each line of text, and it will also prevent you from dwelling on any line for too long.

Over-focused reading: Too detailed and slow, overly focused on specific words or sections. Attention to detail is important, but an awareness of the broader context of what you are reading is also vital to remember.

Unfocused reading: Too shallow and focused only on the bigger picture, without paying much attention to details such as the author's theoretical position, methods or key arguments.

Inattentive reading: Related to being unfocused, this is where you miss key words or do not fully follow the sequence of points made by the author. This can lead to problematic conclusions being reached, based on misunderstandings (Cotterill, 2017).

Figure 13.2 Some common reading problems.

Reading effectively

Now that you understand skimming and scanning and can apply the SQ3R approach to speed-reading, here are some additional points to help you to become an effective academic reader.

Time management

Remember, you cannot read everything, so try to be strategic about your use of time and careful about what you read – and what you do not read (apply the skimming and scanning approach above to help).

Ask questions

As you read, it can be useful to regularly question yourself about the text, the approach of the author, the main argument, how the text compares to others you have read on the topic and how useful the reading could be for your assignments.

Location

Think about where is best for you to read in order to maximise your use of time. Being somewhere quiet, comfortable and warm can help, somewhere free from distractions. At busy times, this may not be the university library.

Time of day

You may be most alert in the morning or afternoon and find it challenging to concentrate in the evening. Being tired can slow down your reading speed. Some people also find it easier to read by daylight rather than by electric lighting.

Take breaks

Remember to take regular breaks, so you do not lose focus (see Chapter 23 on balance and wellbeing).

Underlining, highlighting and asterisking

Marking your text can be useful to help you by identifying the key points or paragraphs relevant to the task that you are working on. It may be that you underline key words or phrases, highlight them or use asterisks or other symbols to note the important parts of the text.

Note-taking and organisation

Keeping organised notes that you will easily understand a few weeks later will help you to continue to develop your critical academic reading skills. It is also crucial to make sure that you keep an accurate list of what you have read to include with your assignment and that you reference your sources appropriately (for some guidance on this, see Chapter 16 on referencing and plagiarism).

WHAT DO STUDENTS SAY?

'I have an app on my phone for the Pomodoro Technique. The app is basically a timer. You tell it how long you want to spend reading and it breaks the time down into a series of equal-length intervals, separated by short breaks. It really helped me to keep focused and motivated while doing my dissertation in the library.'

(Paul, graduated with a BSc (Hons) degree in Geography and Environmental Science)

Summary points

1. Skimming and scanning are useful techniques to employ to help you get a general feel and pick up some key messages so you can decide whether or not you need to read it in more detail.
2. SQ3R (survey, question, read, recall and review) is a useful technique to apply to develop your critical academic reading skills.
3. In order to maintain focus, consider if there are distractions slowing down your reading or your reading is not as focused as it could be.

Key reading

Cottrell, S. (2017) *Critical Thinking Skills: Effective Analysis, Argument and Reflection.* London: Palgrave Macmillan (chapter 9).

Waller, R. & Schultz, D. (2015) *How to Succeed at University in GEES Disciplines: Enhancing Students' Information Literacy Skills.* York: Higher Education Academy. Available at https://www.geography.org.uk/download/ga_conf15_forumhandout2.pdf

CHAPTER 14

WRITING ESSAYS

How to succeed in essay writing

We each remember being very nervous about writing our first essay at university. How much detail should we put in? How much should we read in preparation? Could we use our lecture notes? What if we could not find a key text in the library? Our aim in this chapter is to help you to think about how to succeed in writing essays for your Geography degree.

Essays are a very common form of assessment, particularly in Human Geography. The reason for this is that essays require you to demonstrate a number of different skills: understanding the question; selecting appropriate literature to include; constructing an argument; demonstrating critical engagement with the issue; and reaching a justified and supported conclusion. Essays nearly always have a word limit, which means that to attract the highest marks you need to demonstrate these skills in a concise manner (see Chapter 8 for more guidance on how university assignments are marked).

Over the years, when marking essays the main problems that we have seen students encountering are: not answering the question; having at best a weak structure and, at worst, no structure at all; being too descriptive and not analytical or critical enough (see Chapter 17); engaging only with the core reading and lecture material for a module or course (see Chapter 12); and failing to develop an argument (again, see Chapter 17). Overcoming these potential pitfalls will enable you to maximise your performance in writing essays. That is the aim of this, and subsequent, chapters.

WHAT DO STUDENTS SAY?

'I recall being very frustrated and disappointed when I only got 52% for my first essays when I started university. I always got 'A's when I was at school. I spoke to my tutor and she encouraged me to read more widely and not to use only my lecture notes and key readings. I followed this advice for the next few essays, which meant it took longer and was a lot more work – but I was delighted when I started getting better marks.'

(Tatiana, graduated with a BA (Hons) degree in Geography)

Answering the question

Addressing the question directly is important in writing essays at university. A frequent reason for underperformance is not addressing the question or only partially answering it. Table 14.1 provides examples of some of the key words used in essay questions at university. It is important that you understand what is being asked of you if you are to address the question properly. Many of our students find it useful to keep a copy of the essay question in front of them when writing an essay so that they can constantly refer to it. This helps to ensure that they stay on track and avoid meandering off onto issues that are not of direct relevance to the question.

Essay planning and selecting material

A key challenge in essay writing – and one of the reasons why essays are used as a form of assessment – is that they require you to carefully plan your approach and to select material that is appropriate to the question. There may be a specific lecture – or more likely, a set of lectures – that is particularly relevant to the essay question. That said, to succeed at writing essays at university you must engage with material beyond that which is presented in lectures.

Table 14.1 Key words used in university Geography essay questions.

Essay term	Definition
Analyse	Break an issue into its constituent parts. Look in depth at each part, using supporting arguments and evidence for and against, as well as how these interrelate.
Assess	Weigh up to what extent something is true. Persuade the reader of your argument by citing relevant research, but remember to point out any flaws and counter-arguments as well. Conclude by stating clearly how far you agree with the original proposition.
Clarify	Literally, make something clearer and, where appropriate, simplify it. This could involve, for example, explaining in simpler terms a complex process or theory, or the relationship between two variables.
Comment upon	Pick out the main points on a subject and give your opinion, reinforcing your point of view using logic and referring to relevant evidence, including any wider reading that you have done.
Compare	Identify the similarities and differences between two or more phenomena. Say if any of the shared similarities or differences are more important than others. 'Compare' and 'contrast' will often feature together in an essay question.
Consider	Say what you think and have observed about something. Back up your comments using appropriate evidence from external sources or your own experience. Include any views that are contrary to your own, and how they relate to what you originally thought.
Contrast	Similar to 'compare', but concentrating on the dissimilarities between two or more phenomena, or what sets them apart. Point out any differences that are particularly significant.
Critically evaluate	Give your verdict on the extent to which a statement or finding in a piece of research is true, or to what extent you agree with it. Provide evidence taken from a wide range of sources that both agrees with *and* contradicts an argument. Come to a final conclusion, basing your decision on what you judge to be the most important factors, and justify your choice.
Define	Give in precise terms the meaning of something. Bring to attention any problems posed with the definition and any other interpretations that there may be.
Demonstrate	Show how, with examples to illustrate.
Describe	Provide a detailed explanation of how and why something happens.
Discuss	Essentially, this is a written debate where you are using your skill in reasoning, backed up by carefully selected evidence, to make a case for and against an argument or point out the advantages and disadvantages of a given context. Remember to arrive at a conclusion.
Elaborate	Give in more detail and provide further information on.
Evaluate	(See 'critically evaluate'.)
Examine	Look in close detail and establish the key facts and important issues surrounding a topic. This should be a critical evaluation, and you should try to offer reasons why the facts and issues that you have identified are the most important, as well as to explain the various ways in which they could be construed.
Explain	Clarify a topic by giving a detailed account of how and why it occurs, or what is meant by the use of this term in a particular context. Your writing should have clarity, so that complex procedures or sequences of events can be understood, defining the key terms where appropriate, and be substantiated with relevant research.
Explore	Adopt a questioning approach and consider a variety of viewpoints. Where possible, reconcile opposing views by presenting a final line of argument.
Give an account of	Give a detailed description of something. Not to be confused with 'account for', which asks you not only what, but why, something happened.

Table 14.1 (Continued)

Essay term	Definition
Identify	Determine what the key points to be addressed are, and their implications.
Illustrate	A similar instruction to 'explain', whereby you are asked to show the workings of something, making use of definite examples and statistics, if appropriate, to add weight to your explanation.
Interpret	Demonstrate your understanding of an issue or topic. This can be the use of particular terminology used by an author, or what the findings from a piece of research suggest to you. In the latter, comment on any significant patterns and causal relationships.
Justify	Make a case by providing a body of evidence to support your ideas and points of view. In order to present a balanced argument, consider opinions that may run contrary to your own before stating your conclusion.
Outline	Convey the main points, placing emphasis on global structures and interrelationships rather than minute detail.
Review	Look thoroughly into a subject. This should be a critical assessment and not merely descriptive.
Show how	Present, in a logical order and with reference to relevant evidence, the stages and combination of factors that give rise to something.
State	Specify in clear terms the key aspects pertaining to a topic, without being overly descriptive. Refer to evidence and examples, where appropriate.
Summarise	Give a condensed version, drawing out the main facts, and omit superfluous information. Brief or general examples will normally suffice for this kind of answer.
To what extent	Needs a similar response to questions containing 'How far ...?'. This type of question calls for a thorough assessment of the evidence in presenting your argument. Explore alternative explanations, if there are any.

Source: University of Leicester (2019).

When it comes to planning your essay, it is often best to start by brainstorming and thinking through relevant ideas to include in the essay. You might then read around the topic, being careful to take notes and keep the details of what you have read, as this will help with referencing/citing your sources later (see Chapter 16). Reading around the topic is an important part of doing a Geography degree (see Chapters 4 and 12) and will probably generate even more ideas, which you can add to the list of points that you originally brainstormed. We would then suggest that you come up with an outline structure for your essay. There are lots of ways of doing this, but you may want to think about how you will develop your argument and what structure would be best, so the points follow in a logical order (see below). You can then produce a first rough draft. We suggest that you constantly revisit the essay question while doing so, to ensure that you maintain your focus on the question.

Our experience is that many students stop at this point and submit their work. Instead of doing this, it is best to read through your work again to see if there is anything that you have missed or have said too much about. By this time, you will nearly have a final draft. At this stage, you should check that you are complying with the upper word limit (and cut down the word count, if necessary). Crucially, the final step before submission is where you carefully edit your work to check for typos, grammatical errors, referencing problems and so on.

> ## TOP TIP
>
> Keep a detailed record of the materials you read – including name of author, the year and page number – for use in your essay. You will need these to reference (aka cite) your ideas properly (see Chapter 16).

Thinking about structure

A well-structured essay means that your work will be easy to read, will flow logically from point to point and from paragraph to paragraph and will address the question in an organised and systematic way. To help with this, despite what you may have been told in the past, you will probably need to use subheadings to break your material down into sections. This guides the marker through the essay and helps to organise your ideas into a logical, easy-to-follow order. As we will see below, most essays have at least three sections – one called *Introduction*, one called *Conclusion* and at least one section in between.

> ## TOP TIP
>
> When writing an essay, select material from a range of sources, including beyond the lecture material and key readings.

Introduction

This is your opportunity to say why the issue that your essay is discussing is important – you should outline the issue from your perspective to demonstrate your understanding of it. Your introduction should also include a summary of how you plan to address the issue and the argument that your essay will make (see Chapter 17). When you revise your rough draft, you should double-check to see that you have followed the structure that you outlined in your introductory outline of the essay. The introduction is one of the most important parts of an essay. Remember that first impressions count, and that this is the first part that the marker will read.

WHAT DO STUDENTS SAY?

'Over the course of my Geography degree, I've learnt how important a good introduction is and I think my marks have improved . . . . After all, first impressions count and when your lecturer is marking so many, it can help yours to stand out from the crowd.'

(Jane, graduated with BA (Hons) degree in Geography and Planning)

Main body of the essay

This should be the longest section of the essay (probably around 80 per cent of it) and is where you get to demonstrate your knowledge of the issues and show how well you understand them. It is important to split this section into paragraphs, each with your own contribution to the question being addressed. Depending on the length of your essay, you may consider breaking the main body down into more than one section. We would avoid naming these 'Main body of the essay 1', 'Main body of the essay 2' and so on, in your actual written work. Instead, we suggest that you select subheadings that reflect the points that you are making.

Conclusion

Mirroring the introduction, a well-written conclusion is important and can leave the marker with a very positive final impression of your work. It can be useful to briefly review what you have covered in the essay in relation to the broader debates. You may also want to consider the main arguments and then highlight the most important components of these. This is your last chance to show that you have directly answered the question.

TOP TIP

Have a strong introduction and use the wording of the question in your conclusion. This helps you to explain how you have answered it directly. To assist with organisation and structure, use subheadings to break up your essay into sections – one called Introduction, one called Conclusion and at least one section in between.

Approaches to structure

As well as having an introduction, main body and conclusion, there are several ways to organise the material in your essay, depending on the specific question that you are addressing and what you decide is the best way to respond to this. This is one of the reasons why carefully planning your essay is important. One approach is to adopt a chronological structure, where you explore an issue or a process over time. This approach can enable you to demonstrate your understanding of how a process works or can help you to show that you have understood events over time, and how they have shaped a more recent change or outcome. A similar approach could be to look at various phases of an issue and how these work.

Another approach to structure is to adopt a thematic focus. To do this, you need to identify a set of themes or issues that relate to the question. For example, you may find themes and then explain or explore the key characteristics of each, and how they have shaped the main issue that your essay is focusing on. A related method is to compare and contrast specific issues – this can take many forms, but examples include outlining the arguments for and against, positive or negative, and then evaluating them in relation to the question.

TOP TIP

Consider carefully the various ways in which you could approach the structure of your essay, and select the approach you think is most suitable.

Summary points

1. To succeed with essay writing, you need to answer the question, have a clear structure and read beyond the material that is presented in lectures, seminars and tutorials.
2. To answer an essay question well, it is important to understand its key terms and then to carefully plan your essay and select the appropriate material to include.
3. As well as thinking about your introduction, the main body of the essay and the conclusion, it can be useful to consider the various approaches to structuring your essay, whether chronological, thematic or comparative.

Additional reading

West, H., Malcolm, G., Keywood, S. & Hill, J. (2019) 'Writing a successful essay.' *Journal of Geography in Higher Education*, 43(4), pp. 609–617.

Waller, R. & Schultz, D. (2015) *How to Succeed at University in GEES Disciplines: Enhancing Students' Information Literacy Skills*. York: Higher Education Academy. Available at https://www.geography.org.uk/download/ga_conf15_forumhandout2.pdf

CHAPTER 15

DEVELOPING AN ACADEMIC WRITING STYLE

Getting started

It is one thing to work on your essay-writing skills and to think through your structure and approach (as we did in Chapter 14), yet this will get you only so far unless you are also working on developing an academic writing style. Your academic writing style is likely to improve as you read more for your Geography degree – by thinking about your writing style, and by studying the writing style and approach of the articles you read for your course, you will improve the quality of your writing. Our aim in this chapter is to think about some of the strategies and approaches that you can adopt to enhance your academic writing style as you progress through your studies.

Trevor Day (2013) developed the 'IPACE' approach to academic writing, a mnemonic for Identity, Purpose, Audience, Code and Experience. Hundreds of students have used it. It is good to consider this when working on developing your academic writing skills, rather than diving into an assignment without much planning or consideration of the factors, which can be unhelpful.

> ## TOP TIP
>
> Before starting a piece of writing, consider the issues of identity, purpose, audience, code and experience.

IPACE approach

Identity

Writing is a personal experience, and so it says something about you. You have many identities – you are a Geography student but perhaps you are also a sibling, a part-time worker or a carer, while also being a student and a Geographer. Developing your academic writing style is about your identity and how you bring this to your writing in various ways.

Purpose

An obvious way of looking at purpose is about fulfilling the requirements of the course, module or unit that you are studying. The purpose then is to pass the assignment. However, purpose is about what your aim is for the person reading your work, apart from for yourself.

Audience

The intended audience of your work is likely to be your tutor or lecturer (i.e. the person who will read, grade and provide feedback on your work). It is worth thinking about who they are and how they will be approaching your work. This is about knowing your audience. However, you are also writing for yourself, so you can clearly communicate your ideas.

Code

This is about format, structure and style. Format is about basic factors like the font size, colour and paper layout (e.g. A4). Structure is about the organisation of your work, such as your introduction and conclusion, specific sections of your assignment, and so on (see Chapter 14). Style is about your writing style and how you present the work.

Experience

This is about the knowledge and skills that you bring to your writing, the content you choose to include and how you engage with the process of writing.

WHAT DO STUDENTS SAY?

'I used to write my essays by using my lecture and seminar notes, alongside some reading from the reading list. It took me a while to realise this but, because I wasn't planning my essays and wasn't really thinking through the big picture, I didn't do very well. I started to think things through in a bit more detail before diving in – and, I mean, I really planned things and thinking through the different reasons for the assignment I was doing – and I started to get much better marks.'

(John, graduated with a BSc (Hons) degree in Geography)

Paragraphs, sentences and words

It may be that you feel that you are already good at writing and do not need much help with this. However, to become *really* good at writing takes practice and can make a real difference to your success at university. One way of developing your academic writing style is to think carefully about the paragraphs, sentences and words as you write, as well as while you are editing your work.

It is important that you use paragraphs when you are writing. A paragraph is essentially a 'wallet of information', and it often focuses on a specific point that is then explained or explored in further detail throughout the paragraph itself. Paragraphs should normally have around four to seven sentences. If you notice that a paragraph is getting quite long, it may be that it needs to be split into two paragraphs. See Table 15.1 for the types of sentences that you often find in a paragraph, and their role.

The topic sentence (or topic introducer sentence) is often the first sentence in a paragraph, and is the one upon which all the others rely. The most common approach is to start a paragraph with a topic sentence, then use the rest of the paragraph to support your point, as outlined in Table 15.1. Alternatively, you could do it the other way around, where you make several points and then end

Table 15.1 Types of sentences and their role in a paragraph.

Type of sentence	Role in the paragraph
Topic introducer sentence	Introduces the overall topic of the paragraph (generally in the very first sentence).
Developer sentence	Expands the topic by giving additional information.
Modulator sentence	Acts as a linking sentence, and is often introduced by a signpost word, moving to another aspect of the topic within the same paragraph.
Terminator sentence	Concludes the discussion of a topic within a paragraph, but can also be used as a transition sentence, where it provides a link to the topic of the next paragraph.

Source: McMillan & Weyers (2012, p. 247).

the paragraph with the specific point that you want to make, having provided the evidence and explanation beforehand. Paragraphs are also important to the visual structure and presentation of your work – they guide the reader through, as you build up your argument.

In addition to taking care with paragraphing, it is useful to consider the sentence structure and content. A sentence must have a verb (i.e. a doing or action word) and end with appropriate punctuation. There is a choice: simple, short sentences; compound sentences (where two simple sentences are joined by a word such as 'and' or 'but'); and more complex sentences, which are often longer. It is worth considering what words are placed at the end of a sentence, as these are often seen to be of the greatest significance. It is important to avoid shorthand forms of words, which you may have become accustomed to using on social media, and instead spell them out fully. For example, it is best to avoid 'it's' and 'couldn't', and instead write these as 'it is' and 'could not'.

TOP TIP

Your academic writing should be in formal, continuous prose. However, avoid using big words and long sentences just to try to make yourself look clever. To get good marks, the key things are to show the extent of your learning and that you understand the key ideas – and these can be achieved more easily if both the sentence structure and the words that you use are simple.

The selection of the most appropriate words to use in your writing is important. There will be specific words or phrases that are relevant to the particular field of Geography that you are writing about, whether this be Social and Cultural Geography, environmental research or glaciation. It is important to use these words appropriately and apply them correctly. At the same time, care should be taken not to use overly complex words or phrases. Some students feel the need to use 'big words' to demonstrate their level of intelligence. Often, this is not necessary and can, on occasion, be unhelpful and distracting. There may be specific terminology that you need to use, yet this does not mean that you should become involved in highly complex language.

It is also crucial that your writing is non-discriminatory. You should not use language that is offensive to any specific social group. For example, care should be taken to avoid inadvertently using gender-specific terminology, such as 'mankind' (use 'humankind', instead). You should avoid assuming that phenomena are male by using 'his' or 'him' (instead use 'his/her' or 'them') and, ideally, you should use examples about and from both men and women. The same applies when it comes to writing about issues of disability, race, religion, sexuality, place and so on.

Grammar and punctuation

Good grammar and clear punctuation are crucial, to achieve success in academic writing. These help your effective communication using the written word, and so are a key piece in the jigsaw of your academic success.

Be clear about when and when not to use *commas* – this is important, as commas can completely change the meaning of a sentence. Commas tend to be used to separate out items in a list, to join clauses that include 'and', 'but' or 'so', and to mark off a phrase at the beginning, or in the middle of, a sentence.

Learn when to use a *colon* and how useful they can be to add further illustration or explanation: they follow a complete sentence or remark. The further explanation could be a list of points, a quotation or just extra information on the initial statement. Avoid placing a colon in the middle of a sentence or using a semi-colon when you need to use a colon.

Consider where you could use *semi-colons* to enhance your writing style. These are generally used: to demonstrate a clear connection between two sentences (instead of a full stop); to separate out items in a list; or to replace connecting words, such as 'however' or 'therefore'.

Ensure that you use *apostrophes* appropriately. They are used to indicate possession (e.g. Simon's book) or a missing letter ('you're' instead of 'you are'). Do not use apostrophes in dates, such as 1960s, or in acronyms, such as NGOs.

Tips for succeeding with writing

Writing well is about communicating your ideas clearly. It is also about using the correct spelling, punctuation and grammar. While the points that we have outlined above are important, below are some more tips that will help you to improve your academic writing style.

Write and read regularly

The more often you read and write, the better your writing will become. Writing regularly will help to increase your confidence about writing and also your style and approach to writing.

Get to the point

You should try to get to the point as quickly as possible and avoid talking around the subject in a vague manner that wastes words and misses the point. The notes about paragraph structure, above, will help here.

Avoid writing in the first person

Academic writing – especially in Physical Geography – is scientific and evidence-led, so be careful when writing in the first person. For example, 'The evidence suggests that …' is a much stronger way of phrasing an argument in your essay than 'I believe that …', particularly if you then draw upon your reading to outline what the evidence is. Writing in the first person is more common in Human Geography and can be a useful way of demonstrating your critical thinking and analysis of the topic being explored.

House style

It is useful to check to see if your department or subject area requires you to write in a specific house style. If they do, you should make sure that you follow its guidance.

Avoid clichés and metaphors

Care should be taken with the use of overly detailed language, metaphors and analogies. These can distract from the main point of your writing and are often open to misinterpretation (which is the last thing you want, when you are trying to communicate clearly!). It is best to select language that is suited to the topic – you are not writing a newspaper article or a novel.

Edit your work

You should edit your writing during the process of writing, as well as when you have completed it. Editing can focus on content and format. It is also a useful way of cutting down on needless words and making your points clearly.

Proofread

One of the biggest problems we find when marking students' work is that it has not been properly proofread. Proofreading your work closely and making the necessary adjustments as you go can help your overall writing style and will increase your chance of achieving high marks. Often, it is useful to leave a piece of writing aside for some time before proofreading it. Sometimes, reading a piece of writing out loud can help you to spot any inaccuracies or inconsistencies in the writing style or approach.

WHAT DO STUDENTS SAY?

'I did English Lit. at school and so I thought I'd be okay at writing Geography essays at uni – but the style was different, and I couldn't get the hang of it . . . . In the end, my tutor recommended I went to the university's free Writing Development Centre. A few one-to-one sessions with them really helped me to understand what was expected.'

(Sarah-Jane, graduated with a BA (Hons) degree in Human Geography)

Summary points

1. Remember IPACE – Identity, Purpose, Audience, Code and Experience.
2. Make sure that your work complies with basic grammatical rules and uses paragraphs appropriately.
3. Think through paragraphs, sentences and words as you work on developing your writing style.
4. Consider who your audience is, and what this requires from you.

5. Read and write regularly and consider your writing a work in progress, so you can improve it over time.
6. Remember the importance of editing your work and of proofreading your writing upon its completion.

Key reading

Day. T. (2013) *Success in Academic Writing*. London: Palgrave Macmillan.

CHAPTER 16

PLAGIARISM AND REFERENCING

What is plagiarism?

As we discussed in Chapters 4 and 13, in your university Geography assignments you will be expected to refer to the texts that you have read. As the next chapter explains, you will be asked to contrast or evaluate the various ideas that they contain. As a result, it is inevitable that in many answers you will end up with lots of different ideas, from lots of different texts, mixed in with your own ideas. This usually produces a very good assignment, but you need to be careful to make clear to the marker which ideas and phrases are your own and which you have obtained from the texts that you have read. If you fail to do this, you risk being accused of plagiarism, which is viewed as a form of cheating by all universities' Disciplinary Committees and Boards of Examiners. We have known students who have been expelled from university for serious and deliberate plagiarism, so read this chapter thoroughly to make sure that the same does not happen to you!

The question you are asking, no doubt, is 'what is plagiarism?' The *New Shorter Oxford English Dictionary* (OED, 1993, p. 2231) gives three meanings of the verb 'to plagiarise':

- To take and use as one's own (the thoughts, writings, inventions, etc., of another person).
- To copy (literary work, ideas, etc.) improperly or without acknowledgement.
- To pass off the thoughts, work, etc. (of another person) as one's own.

Oxford Brookes University (2018) uses this definition: 'Presenting or submitting someone else's work (or ideas), intentionally or unintentionally, as your own'.

The key here is to notice the phrase 'intentionally or *unintentionally*'. Colin Neville (2007) helpfully provides a list of the main forms of both intentional and unintentional plagiarism.

Forms of intentional plagiarism

1. *Not adding in-text references (aka citations) in your writing.* Notice what we did a few lines above – we mentioned that this list is the work of Colin Neville. His is a really helpful list, but it is not one that we came up with ourselves. We like Colin Neville's list so much that we are using it here, and to avoid plagiarism we have clearly acknowledged (in what is called an in-text reference or citation) to whom the list belongs. If we had not, we would be cheating by passing off this list as our own. More on this sort of referencing will follow later in this chapter. For now, let's get back to Point 2 on Colin's list!

2. *Omitting sources from your list of references/bibliography.* At the end of every assignment you need to provide a bibliography or reference list – in other words, a list of the texts that you have used in that assignment. In the example above, we wrote 'Colin Neville (2007)' so that you can easily find the full details of the text containing the reference list/bibliography at the end of this book, then go and look it up in your university's library and find the text. As we have provided enough information for you to look up the original list, this makes it absolutely clear to everyone exactly where we got this list.

3. *Omitting the list of references/bibliography completely.* Given that university Geography markers reward you for giving the evidence of wider reading (see Chapters 4 and 13), we are not sure why you would want to do this. Apart from it being a requirement, the reference list/bibliography is a clear demonstration to the marker that you have undertaken wider reading.

4. *Taking an assignment written by another student and submitting it as your own work.* Hopefully, we can all agree that this is plagiarism – and theft! Also, as we will explain below, most universities employ software to detect if the same essay has been submitted previously (either at their own or at another university).

5. *Having someone else write your assignment and then submitting it as your own work.* Clearly, this is cheating!

6. *Cheating in exams.* Don't be tempted.

7. *Purchasing from the internet a completed assignment and submitting it as your own work.* There are at least three problems with this: (a) it is another form of cheating; (b) all university Geography courses are slightly different, depending on the research interests of the staff in the department, and often the essays provided on the internet are quite generic and therefore not very useful; (c) again, as we'll explain below, most universities have software to detect this.

8. *Attempting to gain credit by submitting the same essay for two different assignments.* This usually happens when students are taking two modules on similar topics (e.g. Political Geography and Geopolitics). The content is similar; so, students submit part of the same essay for both assignments. Even if the essay was written by the student, this is plagiarism. As with point 4 above, most universities have software to detect whether parts of the same essay have been submitted twice.

Unintentional forms of plagiarism

1. *Either inaccurate or incomplete in-text references/citations or reference list/bibliography.* As we'll explain later in the chapter, there are some recognised systems for referencing a text to help you to avoid plagiarism. Your Geography department will have a system that it wants you to use throughout your degree – usually it is not a free choice (see later in the chapter for further discussion of the reasons why). Not knowing how to use this referencing system properly in your essays is no excuse – get it wrong, and you risk being accused of plagiarism. Claiming you 'forgot to note down where a quote came from' is not enough for a university to let you off. Helpfully, Maier et al. (2009) provide a handy table to remind us what information we need to note for the various types of texts, if we are to reference them accurately at the end of each assignment.

	Author	Year of publication	Title of publication	Title of article/ chapter	Issue	Place	Publisher	Edition	Page numbers	URL	Date accessed
Book	√	√	√			√	√	√			
Chapter in book	√	√	√	√		√	√	√	√		
Journal article	√	√	√	√	√				√		
Internet	√	√	√							√	√

Source: Maier et al. (2009, p. 305).

2. *Poor quoting and/or poor paraphrasing.* In our experience, this is the biggest cause of unintentional plagiarism by students – yet the rule is simple. If you do not change any words at all in an extract from a text, it is a quote, so it needs quotation marks. If you do not wish to quote something but just to mention it, then you need to change the wording/phrasing substantially, so that you are completely rewriting the same idea in your own words (i.e. paraphrasing). This is the part that students often do not do thoroughly enough. Often, to speed up the assignment-writing process, instead they just change or add an occasional word and then claim – by omitting to use quotation marks – that it is their own phrase. But it is not. It is still basically a quote, just one with one or two words added or changed.

Whether you use a quote in quotation marks or turn it into an idea that you have entirely rephrased in your own words, both need to be referenced in specific ways – but we'll get to that in a moment!

> ## TOP TIP
>
> While at most universities the punishment for intentional plagiarism is harsher than unintentional plagiarism, both types carry severe penalties, as students are expected to know what plagiarism is and how to avoid it. Plagiarism does not just apply to words. It can apply to pictures, graphs, diagrams, websites and so on.

How do lecturers spot plagiarism?

It is not as difficult as you might think for academics to spot plagiarism. The first thing to remember is that university Geography class sizes tend to be fairly large, so the lecturers are marking the same assignment very many times. One consequence is that we quickly become very familiar with the texts that most students use, and we have probably already read them ourselves and used them in our own research. Sometimes, if we believe that we recognise a section of an essay, we can take just a few minutes to search online for one or two phrases to find out if it is really a quote that has not been referenced. It is even easier to spot plagiarism when two students have used the same quote and, while one has referenced it, the other has not.

Another tell-tale warning sign is an essay in which the writing style changes frequently. This is normally a sign that the student is copying and pasting chunks of text from various sources without referencing them, and then making the plagiarised text flow better as an essay by adding just a few words of their own between the chunks.

Finally, most universities have electronic plagiarism-checking software, such as Turnitin. An assignment is submitted electronically by a student, at which point the software scans it for possible plagiarism and sends a report to the marker. Once submitted, the assignment is stored and used to check for plagiarism in later students' work. Identifying any matching text in the two assignments prevents a student from passing off another's as their own original work. The systems are international, so the students do not even have to be at the same university.

TOP TIP

In your assignments, don't use too many quotes from texts, even if you reference them correctly. If there are too many, it is difficult to demonstrate to the marker that you understand the material, as you are just cutting and pasting other people's ideas. Paraphrasing shows that you have thought about what the texts you are using have to say and that you have some ideas of your own to add. Of course, both the quoted and the paraphrased texts must be referenced correctly.

Referencing

Hopefully, it is clear from the previous section that you need to reference a source every time that you make a point or refer to data and other information that are someone else's work. To help with this, there are various referencing styles in common use. They are all very different in terms of how they set out the information but, in essence, they all have the same aim: to allow the reader to identify the original source that the writer has used, thus avoiding plagiarism.

Each Geography department will have its own preferred referencing format, as does every journal or book publisher. One common mistake that new students make is to read several texts with contrasting referencing styles, fail to recognise this and, in their first assignment, use some aspects of each in a mangled, hybrid style of referencing. Please don't do this! It is vital that you find out in the early weeks of term which referencing style your department uses and learn how to use it. If you use the wrong one in your assignments, you should expect to be penalised.

> ## TOP TIP
>
> Before you submit any assignments, find out which referencing system your Geography department expects you to use and learn how it is formatted. Not all departments use the Harvard system of referencing, yet most university libraries do have a 'How to ...' referencing guide that you can pick up. Googling the name of the referencing system that your department uses will help you to find a guide.

Using the Harvard system to reference sources within the text of your assignment

Of all of the referencing styles, the Harvard referencing system (aka the author-date system) is probably the one that is usually used in Geography, and so it is this approach to referencing that the rest of this chapter focuses upon.

The Harvard system includes in the text only brief details of the source document that is being used. The full details of the source are then given in a reference list or bibliography at the end of your assignment. This allows the writer to acknowledge their sources adequately in the text of their assignment without interrupting the flow of the writing too much. It is this Harvard style that we are using to reference source documents throughout this book.

A Harvard-style reference in the text of an assignment includes the surname of the author(s) and the year of the publication. This information is included in brackets at the most appropriate point in the text – usually after you have referred to someone else's idea, often at the end of the sentence. It should always be before the full stop. For example, this next, made-up in-text reference indicates to the reader that the point being made is drawing on an idea from a (fictitious) document written by Ian Smith and Paul Jones, published in 1991.

> The willingness of students to learn depends greatly on their study skills (Smith and Jones, 1991).

An alternative layout of the same in-text reference, which you might use to help with the flow of your essay, is shown in the example below:

> Smith and Jones (1991) believe that the willingness of students to learn depends greatly on their study skills.

Alternatively, you could use this layout, where the reference appears in the middle of the sentence:

> As demonstrated by Smith and Jones (1991), the willingness of students to learn depends greatly on their study skills.

When you include a direct (i.e. word-for-word) quote from another author's work, in addition to the information in the above examples you should use quotation marks around the text that you have copied and always add the page number – either after a 'p.' or a colon. For example:

> As demonstrated by Smith and Jones (1991, p. 347), 'study skills are a key determinant of the success of students undertaking Geography degree courses'.

Or

> As demonstrated by Smith and Jones (1991: 347), 'study skills are a key determinant of the success of students undertaking Geography degree courses'.

An alternative referencing layout, which again might help with the flow of your essay, is:

> 'Study skills are a key determinant of the success of students undertaking Geography degree courses' (Smith and Jones, 1991, p. 347).

Or

> 'Study skills are a key determinant of the success of students undertaking Geography degree courses' (Smith and Jones, 1991: 347).

When a publication has two authors, give both surnames and the year, e.g. (Smith and Jones, 1991). Put the surnames in the same order as they appear on the text that you are reading. When a publication has three or more authors, it is usual to give the surname of the first author then 'et al.', which is Latin for 'and the others': for example, Brough et al. (2010).

Do not forget that you should include a reference to the source of any table of data, diagram or map that you use in your assignment. If you have used one of these, it is usual to add a reference to the end of the caption to explain what the table, diagram or map shows. As you will normally be copying tables, diagrams and maps identically from the original source, you will usually need to include the page number of the original in your reference. An example of a caption followed by the reference is:

Table 1: Factors influencing the success of students studying Geography (Smith and Jones, 1991, p. 12)

Or

Table 1: Factors influencing the success of students studying Geography (Smith and Jones, 1991: 12)

WHAT DO STUDENTS SAY?

'Referencing correctly took me ages, at first. I used to sit with the library's Harvard Guide pinned to the wall behind my laptop screen, and I had to look up at it every time I wanted to reference anything. It was probably sometime in second year when I realised that I was hardly using it for most assessments. Websites like CiteThisForMe or Endnote can help, but it is definitely best if you learn how to reference yourself.'

(Katrina, graduated with a BA (Hons) degree in Geography)

How to format a Harvard-style reference list or bibliography at the end of your assignment

When using the Harvard system of referencing, all the sources that you use in the text of your assignment must have an accompanying statement, located at the end of your assignment, showing their full publication details in an alphabetical list of author surnames. This is your reference list, or bibliography. Importantly, there should be only one list – don't sub-divide it by the type of source (books, journal articles, etc.). There is not much room for flexibility, as all of the sources that you have used must be listed and each must be formatted correctly, consistently and accurately.

Unhelpfully, for the various types of sources there are slight variations in how you format the reference and in which details you need to give. The examples in Figure 16.1 indicate the main principles, as you need to get this right.

From time to time, there may be other types of sources that you wish to reference, for example newspapers and websites. However, our best advice is first to check with your lecturer that these are good sources to use for the assignment that they have set you – in most cases, they won't be! Only once your lecturer has given you the green light to use them should you explore how to reference these in-text and as part of your reference list/bibliography.

WHAT DO STUDENTS SAY?

'Don't use your school textbooks, Wikipedia or other web pages in your assignments. When lecturers talk about "using a range of sources", they mean we have to use specialist texts, written by other academics – like books and journal articles.'

(Safwah, graduated with a BSc (Hons) degree in Geography)

How to reference a book in your reference list/bibliography

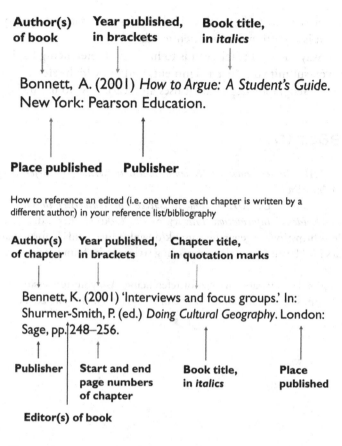

Author(s) of book **Year published,** in brackets **Book title,** in *italics*

Bonnett, A. (2001) *How to Argue: A Student's Guide.* New York: Pearson Education.

Place published **Publisher**

How to reference an edited (i.e. one where each chapter is written by a different author) in your reference list/bibliography

Author(s) of chapter **Year published,** in brackets **Chapter title,** in quotation marks

Bennett, K. (2001) 'Interviews and focus groups.' In: Shurmer-Smith, P. (ed.) *Doing Cultural Geography.* London: Sage, pp. 248–256.

Publisher **Start and end page numbers of chapter** **Book title,** in *italics* **Place published**

Editor(s) of book

How to reference a journal article in your reference list/bibliography

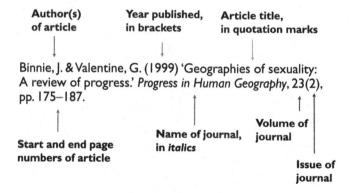

Author(s) of article **Year published,** in brackets **Article title,** in quotation marks

Binnie, J. & Valentine, G. (1999) 'Geographies of sexuality: A review of progress.' *Progress in Human Geography,* 23(2), pp. 175–187.

Start and end page numbers of article **Name of journal,** in *italics* **Volume of journal**

Issue of journal

Figure 16.1 Formatting a Harvard-style reference list or bibliography at the end of your assignment.

Summary points

1. Plagiarism is a form of cheating. Universities treat plagiarism very seriously, whether it is intentional or unintentional.
2. The best way to avoid plagiarism is to have good referencing habits.
3. Before you submit your first assignment, find out which referencing style your university Geography department uses and learn how to use it.

Key reading

Deane, M. (2010) *Inside Track to Academic Research, Writing & Referencing*. Harlow: Pearson Education.

Waller, R. & Schultz, D. (2015) *How to Succeed at University in GEES Disciplines: Enhancing Students' Information Literacy Skills*. York: Higher Education Academy. Available at https://www.geography.org.uk/download/ga_conf15_forumhandout2.pdf

Useful guides for Harvard and other referencing styles can be found at www.citethisforme.com/guides

Endnote is a popular software package for referencing. Ask your university library if it has a subscription, so that you can use it free of charge: www.endnote.co.uk.

CHAPTER 17

ARGUING AND THINKING CRITICALLY

Moving from being descriptive to being critical

If memorising key facts, dates and events was the key to getting a good mark at school or college, you may have performed well in Geography examinations or coursework. Partly because of this, Geography students at university fall easily into the trap of writing descriptive answers, where they are essentially recalling 'facts' or replicating information as it stands, without any form of analysis or critique. This might include describing the environment, recording measurements, noting dates or times or giving a chronological account of events. You may write well and clearly and may reference your sources properly (see Chapter 16), but just describing an issue is unlikely to secure high marks (see Chapter 8). Descriptive writing is quite easy and it can use up a great many words, too.

Although to succeed at university it is useful to be able to recall key pieces of information, thinking critically will allow you to excel. Often, for marks of 60 per cent and higher, the criteria to assess your work will include 'thinking critically', 'critical engagement with' and 'critical analysis of' course materials (see Chapter 8). We are not saying that you should avoid description; some is necessary to demonstrate your knowledge and understanding. What we are saying is that this needs to be balanced by critical and analytical insights into the issue that you are exploring.

TOP TIP

Review what you are writing and ask yourself if it is simply a descriptive account or if you are making critical or analytical points about the issue at hand.

What is critical thinking?

So, what does it mean to think critically? Thinking critically is not about being negative, or about pointing out all the weaknesses or flaws in an issue. Furthermore, being critical is not necessarily about ensuring a balanced presentation or the pros and cons, strengths and weaknesses. In order to move towards a more critical frame of thinking, ask yourself the *what, why, who, how, when* and *where* questions about the topic, issue or question that you are seeking to address. Being critical in your assignments at university is about asking these questions (and others), thinking things through carefully and not relying on 'common-sense' explanations or understandings. Thinking critically can also involve thinking about the topic that you are working with in such a way that it helps you to identify the key literature and any issues that might not have been identified if you had simply taken the topic at face value.

In a sense, you already exercise critical thinking in your everyday life if, for example, your friend recommends a new gym that has some great membership deals; if the supermarket has a new offer on your favourite crisps; or if your mother suggests that you should wear red more often, as it suits you. You would probably evaluate each of these, work out which you agree and disagree with and take some of them up but avoid others. You would consider each offer and suggestion. Next, you would make judgements and appraisals about them, and then move forward after making up your own mind. This is critical thinking.

Carrying these skills through to your degree is the key to constructing coherent arguments and thinking critically. You have already learnt to be selective – for example, you have chosen to study a particular course at university, having evaluated it against others. Even if you are not studying at your first-choice university, you have still (hopefully!) made the considered and carefully evaluated decision to study Geography at university. You may have been encouraged to study another subject or had a teacher or family member persuade you to do something else yet, in the end, you chose to study Geography. The chances are that, in doing so, you

WHAT DO STUDENTS SAY?

'At sixth-form college we were always told to present a balanced argument in our essays, to give the pros and cons of everything. It wasn't until well into second year that I realised that this wasn't enough and that being critical actually means not just accepting what authors wrote at face value and summarising it. Instead, being critical means questioning and probing their ideas, whether they can be seen as pros or as cons.'

(Usman, graduated with a BA (Hons) degree in Geography and Planning)

were thinking critically when you evaluated and weighed up the evidence and the various viewpoints before making your decision. If we asked you to talk us through your thought processes, you would be making an argument about why studying Geography at university is right for you.

To help us think about this in another way, the Open University has created a 'stairway to critical thinking' (see Figure 17.1). This can be a useful way to understand critical thinking. In the diagram, the steps at the bottom provide support for the more advanced, higher-level steps nearer the top – as you climb, you move towards using higher-level skills, such as being more critical.

Another important part of being critical is deciding what you do and do not choose to read, being judicious about the texts on your lecturer or tutor's reading lists and the extra books from your university's library that you choose to engage with (see also Chapters 4 and 12). This is about evaluating the credibility of your sources and making considered decisions about what to read – and what not to read – for a specific assignment. The key to being critical is to evaluate the quality of your sources. For example, in what year was the article you have found written? Who is the author? Is it a source from a reputable university or organisation? The abstract of an article can be useful to consult as an overview, without reading the whole text.

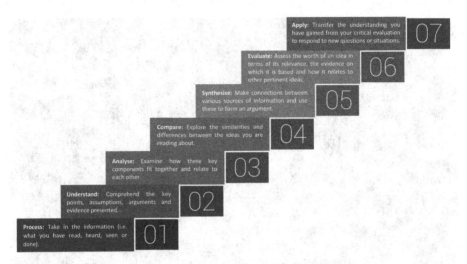

Figure 17.1 Stairway to critical thinking.

Source: Adapted from Open University (2013, p. 12).

Developing critical thinking skills is important in terms of attracting the highest marks in university assignments (see Chapter 8). However, they are also increasingly valued outside of university. Many employers look for graduates who can think critically and engage with issues from diverse perspectives. Critical thinking also helps with presentation skills and promotes creativity.

WHAT DO STUDENTS SAY?

'I love my new job in advertising, as it's all about being able to think critically and outside of the box. I have my Geography degree to thank for that!'

(Khadija, graduated with a BSc (Hons) degree in Geography)

Making a coherent argument

Critical thinking is a key mechanism for constructing an argument in a persuasive way. A clear writing style (see Chapter 15) and evidence of your critical thinking by appropriate reference to your sources (see Chapter 16) provide the foundations for a coherent argument. Part of being critical is about being selective about which points you make and the order in which you make them – you cannot refer to everything and you probably cannot say everything that you want to, either. Being critical is also about providing evidence for the statements that you make, so to justify your argument you should ensure that you reference your sources. Setting up your argument in the introduction and returning to it in the conclusion helps to ensure clarity (see Chapter 14).

If the argument you want to make is coherent, it should be possible to outline it in one sentence and explain it to someone else easily. Sometimes, you may find that you are working with more than one argument and, in such cases, it is useful to develop a central argument, with other points being subsidiary. Being focused and specific can help to ensure that your argument is not too broad or lacking in focus.

Some common issues that restrict critical thinking and limit an argument's coherence include: exaggeration or oversimplification of an issue; over-generalised points that lack specifics or focus; reinforcement of problematic stereotypes; over-reliance upon one or two specific sources of evidence; and a reliance on problematic sources, such as politically biased, non-academic sources or tabloid newspapers.

TOP TIP

When you have finished your essay and before you proofread it, review it paying specific attention to the balance between description and analysis and to the consistency of your argument.

Summary points

1. Critical thinking is an important skill to develop as you progress through your studies at university. Most marking criteria require evidence of critical thinking skills.
2. As a strategy for helping you to think critically, ask yourself the what, why, who, how, when and where questions about the topic, issue or question that you are working with.
3. Developing critical skills during your degree is partly about making critical decisions on what you choose to read or not read.
4. An important step in developing a critical approach to your studies is making a coherent argument and following this through in your work.

Key reading

Bonnett, A. (2011) *How to Argue*. Harlow: Pearson Education.

Waller, R. & Schultz, D. (2015) *How to Succeed at University in GEES Disciplines: Enhancing Students' Information Literacy Skills*. York: Higher Education Academy. Available at https://www.geography.org.uk/download/ga_conf15_forumhandout2.pdf

Open University (2008) *Thinking Critically*, http://www.openuniversity.edu/sites/ www. openuniversity.edu/files/brochures/Critical-thinking-Open-University.pdf

CHAPTER 18

SURVIVING EXAMS

Types of exam

Along with essays, exams are one of the main methods that universities use to assess Geography students. When preparing for an examination, one of the first points to consider is what form it will take. Whereas at school or college you will have mainly taken unseen written exams, at university they come in many formats. Some of the types that you may need to sit during your Geography degree are listed in Table 18.1.

Preparing for exam success

There are several factors to consider when preparing for exam success. Each of the points we discuss here is worth giving some thought to. Some of the points below might be quite easy for you to address, but others might take more work and will require perseverance. The type of exam you will be sitting should shape how you approach your preparation.

Devise a study plan

A key step in preparing for exam success is to have a clear study or revision plan and to stick to it. When devising a plan, it can be useful to consider your other commitments, such as any part-time work, caring responsibilities, the sports teams that you play for and so on, and to factor these into your timetable. Often it is a mistake to cut out all leisure activities while you revise. Instead, try to include time for relaxation. Remember, sleep is crucial to your functioning to the best of your ability.

Table 18.1 Types of university Geography exams.

Exam type	Explanation
Multiple-choice	These tend to include a diverse range of questions where there is one correct answer to each. Often, these exams are conducted online yet under examination conditions.
Short-answer	These are about demonstrating your knowledge and understanding of specific issues by offering brief explanations or demonstrating your understanding of the issues being asked about.
Essay-based	One of the most common forms of exams used in Geography degrees (particularly in Human Geography), these ask you to answer one or more essay question(s) within a set time. Commonly, essay-based exams last 2 to 3 hours.
Open-book	These often take place under exam conditions, but you can take notes, books and other resources into the exam room with you.
Take-home	Similar to open-book exams, you are set specific questions and have a limited time to address specific questions or tasks, but you can take the exam questions 'home' rather than having to write your answers in an exam room. In reality, students normally answer these exams in the university library.
Unseen or closed-book	The most traditional form of exam, where the questions are only released at the start of the exam and need to be addressed in a limited time, under exam conditions, without using any additional resources.
Seen	The exam questions are released to students in advance of the exam date. Although the exam itself is likely to be under exam conditions, you will have had an opportunity to review the questions and plan your answers.
Data-response	Students are provided with a dataset to study in advance and then asked to respond to questions in a traditional, unseen exam format.

Consider your exam technique

Some students assume that success in exams is achieved by memorising facts, dates and references. A careful approach to revision is clearly important; however, this could be wasted if you do not pay any attention to the issue of exam technique. Put differently, you may have revised very thoroughly and clearly be on top of your material, but if you do not have a clear plan for how to approach answering the questions, you may not end up using the material that you have remembered to best effect. Adopting a critical approach (see Chapter 17) is often better than regurgitating everything that you can remember about the topic.

Look at past papers

Looking at previous exam papers for the module that you are studying and writing practice answers can help you when it comes to exam technique and achieving exam success. Past papers may be available on the website of your Geography department or from your university's library, or you could ask your tutor or lecturer to give you some. For new modules or courses, you could ask the course leader for a mock exam to give you a flavour of the type and format of the questions.

Think about how you will plan your answers in the exam

Taking the time to plan how you will answer questions is a crucial step in the journey to exam success. You could use a mind-map approach, a flowchart or simply a list of section headings or paragraph topics. When revising, it can be useful to plan answers to various types of questions so that you have a strategy to address the questions that might arise.

Practise answering the question

One sure-fire way of not performing as well as you can in an exam is not to answer the question. This is easier than you would think – under pressure in the exam hall, it can be very easy to misread a question or to panic and begin writing too quickly, without careful planning. Once underway, it can be easy to wander off course and write an answer that – while being a strong essay, in and of itself – does not directly address the question that was asked. Again, planning answers to past-paper questions as part of your revision can help you to refine your technique.

> ## TOP TIP
>
> Don't just write down everything you know about each topic in your exam answers – and don't try to shoehorn information that isn't relevant into your answers. A more critical approach is often the key to achieving success in exams (see Chapter 17).

On the day

Aside from the hard work involved in preparing for an examination, it is also important to consider what you do on the day of the exam and the night before. Make sure that you have had plenty of sleep and eat a proper breakfast, as this will ensure that you are as refreshed as possible and have enough energy to concentrate for the full duration of the exam. You may also want to consider how hydrated you would like to be – you want to make sure that you are not dehydrated yet, ideally, you do not want to be running to the bathroom, either.

It may be useful to consider what you do before the exam and who you speak to – or not – on the day. You may find it helpful to meet up with people on your course or module to talk about what you have been revising, how you are approaching the exam or to clarify key points about specific theories or approaches that you have been reading about. However, it could also be that this would work against you, and that it would be best to head straight to the exam room so that you can maintain your focus. Only you will know what works best for you.

WHAT DO STUDENTS SAY?

'How well I did in exams was often about how good or bad I felt on the day. I realised towards the end of my degree that some of my mates were competitive and it really put me off my stride if I spoke to them before exams. So, I deliberately avoided them before exams and found I performed much better.'

(Carl, graduated with a BSc (Hons) degree in Environmental Sciences)

In the exam room

In the exam room, remember your strategy and plan how you want to approach the exam. At the start of the exam, write this down on your script as a reminder to yourself. Indeed, it may be useful to make additional notes about things on your script, if you think that this will help to clear your mind of any specific points that you have been mulling over or struggling to remember.

TOP TIP

Plan exam answers thoroughly. Any notes you cross out before the end of an exam will not be marked.

At the start of the exam, take some time to read all the questions carefully and note down what questions you would like to answer, making sure you are addressing the exam requirements. For example, is the exam asking for two essays to be addressed in two hours, or three answers in one hour? Or, are there several sections of the exam paper, with each section requiring one or more answers?

It is best to plan your exam answers, either planning them all at the start, or planning and immediately writing each question in turn as the exam progresses.

1. Try your best to enter the exam room calmly. You may want to use some of the time before the exam begins to do some simple relaxation and breathing exercises to calm your nerves.

2. Begin by very carefully checking the instructions on the exam paper, highlighting or underlining the key points, e.g. key words in the question, word limits, and so on.

3. Consider the amount of time that you have, decide how best to distribute this to the various sections of the exam and aim to stick to your timings. Three average answers will usually score more marks overall than two brilliant ones and an unfinished third.

4. Where there is a choice of questions, do not start writing immediately. Take some time to consider the potential of each option before making your decision. If you freeze up, take a few deep breaths, re-read the questions and then do your best to proceed with the exam. (Read on for advice on dealing with panic during an exam, if you think this could come in useful.)

5. Once you have made your choice, read the question(s) thoroughly, then re-read to make sure that you have understood and have not made assumptions.

6. With an essay-based question, plan your answer briefly to ensure a strong, critical argument. Keep this simple: no more than section headings and your basic points and examples. If it is relevant, you may find it useful to quickly jot down any sources or quotes to refer to.

7. Remember, you do not have to answer questions in the order that they appear. Some people may want to start by getting the more difficult questions out of the way, while others prefer to build confidence with easier questions first.

8. Throughout the exam, try to stay hydrated.

9. If possible, take regular 'micro-breaks', e.g. a brief pause at the end of writing a paragraph. Try putting down your pen and stopping to think for a moment, which can help you to assume control and collect your thoughts.

10. Towards the end of your exam, try your best to conclude your essays in some way, and find a little time to double-check your answers, if you can.

Figure 18.1 Top tips for sitting exams.

Source: UCL (2019).

The approach that you take is down to your own personal preference. It can be useful to consider writing the answer that you are most confident with first. This can help to build your confidence and self-esteem as you proceed. The risk of doing the opposite – starting with the hardest one – is that you get bogged down with a challenging and unfamiliar question and end up going off course or losing your cool. This can risk upsetting your planned approach, for example if you run out of time towards the end because you spent too long answering a difficult question at the start.

Some tips for success

There are many reasons why students do not perform as well as they could, or as well as they expect. Being aware of the possible pitfalls can help you succeed, as you will be alert to these and thus more able to avoid them. Here we focus on some of the key issues with exams so that you can plan for success.

Recall information effectively

One of the keys to success is being able to recall key arguments, points, authors and so on. Some find it easy to recall information like this. For others, it takes more work. Consider what works best for you, so that you have a clear plan in mind.

Tightly manage your time

Poor time management can create problems, as it may lead to your writing a very long and detailed answer to one question but only a very short answer to another, missing out key points. Be strict about your timing and allocate an equal amount to questions of equal value.

Clear structure and clear presentation

As with all forms of assessment, clear presentation and a sound structure can help you succeed. Poor handwriting can hold you back, especially if the marker is unable to read it. Likewise, a clear structure that leads the reader through your questions always helps, whereas a wandering answer without much structure may prevent you from securing high marks.

Reference your sources

A key mechanism for demonstrating your knowledge of the subject is to reference your sources. This can be challenging in exams – a perfect in-text referencing technique is not normally expected – but you have to try your best (see Chapter 16 for more guidance on referencing).

Give critical insights

Just as critical thinking skills are important at university (see Chapter 17), demonstrating critical insight in exam answers is one route to exam success. Descriptive answers that simply recall key facts may secure a reasonable mark, but such answers are unlikely to be awarded a first-class grade (see Chapters 4 and 8 on approaches to teaching and assessing university Geography).

Use positive self-talk

Maintaining positive self-talk and focusing on positive beliefs (rather than those that limit or restrict you) are important to exam success. A positive approach can help to keep your confidence levels high and make you feel assured about your progress in the exam.

> ## TOP TIP
>
> Ensure that you are answering the question that is asked, rather than addressing another issue or question. Consider using the wording of the question to make sure that you are addressing it.

Feedback on exams

As you are now studying Geography at university, you probably already have experience of sitting exams and are mastering this form of assessment. However, some students find exams challenging and prefer modules with alternative forms of assessment. One of the issues with exams is that they are seen as a form of assessment that does not give you feedback other than a final mark (such as a grade or a percentage mark) or a pass/fail classification.

One way of improving your performance in exams is to establish what works for you and what does not. To do this, getting some feedback on your performance can be very helpful. You may sometimes be offered the opportunity to discuss the exam with your module leader or personal tutor (see Chapter 7 for more guidance on the role of your personal tutor) and to look over your answers and any feedback. If this is not offered automatically, ask your module leader or tutor if you can book an appointment to have a chat about this – often at university you need to proactively ask for the help and support you need.

WHAT DO STUDENTS SAY?

'I was finding that I never did nearly as well in exams as I did with my coursework. I would regularly get 60s for essays and other coursework but my exams were all in the 50s. I met with my tutor and looked over some of my previous exam papers; I could see that my answers were not planned as well as they could be and that I wasn't referencing the literature in my answers. I changed this in my next exam and ended up getting a 73!'

(Gemma, graduated with a BA (Hons) degree in Human Geography)

Summary points

1. Exams can take various formats, including: multiple-choice, short-answer, essay-based, open-book, take-home, unseen or closed book, seen and data response.
2. Preparing carefully for exams is an important step to success. A clear study plan, a focus on exam technique and, by looking at past papers, making sure that you answer the question are all part of this preparation.
3. Alongside carefully planning, think carefully about what you will do on the day of an exam and how you will approach it when you are in the exam room.
4. Remember to reference your sources, and work on developing a critical style in your answers while maintaining positive self-talk.
5. Where appropriate, ask for feedback on your exam performance so that you can assess what is working for you.

CHAPTER 19

DELIVERING PRESENTATIONS

Getting started

Many Geography degree courses require you to participate in presentations, either individually or as part of a group. The presentations may be informal, as part of a tutorial or seminar, or they may be more formal and assessed as part of your overall mark. Developing your presentation skills is important, particularly when it comes to securing employment after graduation. As part of the interview process, many employers ask candidates to deliver a presentation, so having the experience at university is a massive help.

Consider if there is someone you know or have heard speak who is an inspirational presenter – for example, a teacher, a lecturer, someone who has given a TED talk or who has a YouTube channel. Perhaps there are specific points about their style or approach that you could adopt that would help you to succeed in delivering effective presentations. Consider what their slides look like, how many slides they have, how they maintain the attention of the audience and what it was about them that drew you to their presentation style. Whoever you have in mind, no doubt they make presenting look easy. That is the sign of a good presenter, but we need to be aware that in nearly all cases a great deal of practice has gone into making their presentation sound natural.

Where will you present and what facilities are available?

How much time do you have?

Who is your audience?

How will you introduce yourself and the topic?

How will you structure your presentation?

Will you use any visual aids?

How will you keep the audience engaged?

Figure 19.1 Questions to consider when preparing your presentation.

Preparing the content of your presentation

Before starting any preparation, it is useful to think about why you are being asked to deliver a presentation. Understanding the purpose of your presentation will help you to succeed, as you will be able to ensure that you deliver what is required. The purposes of a presentation could include: demonstrating your knowledge and understanding of the course material, such as a key reading or concept; explaining new data that you have collected or analysed; sharing the findings of fieldwork or a desk-based study; or pitching an idea or proposal.

Irrespective of the audience and purpose of your presentation, it is important to be prepared. One important aspect to think about is its content. Like a good university essay (see Chapter 14), a successful presentation is tightly structured and leads the audience through the stages of the material. Consequently, (if you are using PowerPoint or similar) it is useful to have a slide near the start of your presentation to outline its structure and another at the end to act as a conclusion, reviewing the main points again for your audience. In order to put these together, you will need a clear idea of the main points you want to make and to have pulled together useful supporting material and references to literature (where relevant).

Another of the main ingredients in a successful presentation is to have practised what you will say and how it will work. Most presentations have a strict time limit, and this is your guide to how much content to include in the middle part of your presentation – in other words, the bit between your introduction and conclusion slides. How many main points do you need to make to fill the time? How much detail do you need to go into on each one? It may be useful to practise several times to ensure that your delivery is as clear as it can be, that you have the right amount of content and that your slides, handouts or any other visuals align with what you say.

TOP TIP

Speaking for an allocated amount of time is a skill that is acquired with time, and so most markers will not allow presentations to overrun. Make sure you practise your presentation in advance several times. Ideally, practise in the same room that you will be in when you deliver the final presentation.

Visual aids, notes and handouts

One of the factors that can determine the success of a presentation is effective use of visual aids, notes and handouts. An effective visual aid – be it a diagram, a figure, an image or a short video – can add interest to a presentation. It can often convey much more information than simply using words. When it comes to developing slides for your presentation, some recommend the 5-5-5 rule: no more than five words per line of text; no more than five lines of text per slide; and no more than five text-dominant slides in a row.

In addition to any slides as visual aids, you should consider whether you want to have a handout or notes to share with the audience. You may want to use visual aids other than slides. At the same time, you should not bombard your audience with too many types and forms of information, so there is a balance to be struck.

- Keep them simple.
- Limit the number of bullet points and the amount of text you use.
- Limit the use of animation.
- Consider the ordering of your slides.
- Ensure high-quality graphics, charts and images that are readable from the back of the room.
- Keep the background of your slide simple.
- Ensure the text is clearly visible from the back of the room, given the background colour.
- Develop your own style, rather than relying on a template.
- Use colour effectively.
- Choose your font and font size with care.
- Consider using video or audio.
- Explain any shorthand terms or acronyms.
- To avoid plagiarism, remember both to acknowledge your sources when speaking and to reference the sources you used on slides, handouts and notes in the same way as in an essay (see Chapter 16).

Figure 19.2 Top tips for creating slides to accompany a presentation.

Preparing yourself to present

Next, think about how you will demonstrate a positive and calm demeanour during the presentation. One step in the right direction is to familiarise yourself with the room in which you will be presenting and any IT equipment that you will be using, such as PowerPoint slides. The room that has been booked for the presentation session may be one that you already know well from a previous class, or it may be new to you. Either way, it is unlikely that you will have stood at the front before. Doing this helps you to become familiar with the space and can help to settle any nerves. You could consider checking out the room in advance or turning up early so that you can assess the space and how you will use it.

WHAT DO STUDENTS SAY?

'I was so nervous giving my first presentation that my notes ended up being effectively a speech that I read out word for word! This was a mistake, as I read really fast and didn't look up at the audience once. It took another couple of presentations and A LOT of practice before I managed to get to a stage where my notes were only a few bullet points as prompts and I just spoke naturally, like having a chat with the audience. Only then did my marks massively improve.'

(Jo, graduated with a BSc (Hons) degree in Physical Geography)

Positive visualisation can help you to achieve a successful outcome. Imagine delivering a successful presentation and receiving positive feedback from your tutor or seminar leader. What would that look like and feel like? Before going to the room where you will present, you could do some of the warm-up techniques that actors use before going on stage. For example, practise a 'power stance' to boost your confidence. Then do a few circular shoulder shrugs, both forward and back. Next, let your arms hang loosely by your sides and give them a good shake. Lastly, rotate your head in a clockwise direction, so you are looking at the ceiling, the wall to your right, the floor and the wall to your left. Then do the same again in an anti-clockwise direction. These exercises help to take the tension out of your body and make you feel relaxed. As you enter the room, taking deep breaths and smiling can help, too.

TOP TIP

When presenting, it can be useful to change your pace at key points and to pause between sections, so that your audience can process the content of your presentation. In addition, pay close attention to your body language – eye contact with your audience and open, positive body language are vital if your audience is to engage with what you are saying. Don't fold your arms, look down or sit down!

Group presentations

Many of the general points in this chapter about presenting as an individual also apply to group presentations. However, given the collective nature of a group presentation, careful planning is crucial. It is often best to work on the presentation as a group and to plan the order, outline and content collectively rather than individually. It may be useful to evaluate the strengths of specific team members and where best to deploy them in the presentation – perhaps one person in the group would be best to do the confident introduction, whereas others might explain well some of the detail of the data that you have collected, while yet others may have a good grasp of a key concept that you are working with.

With group presentations, the introduction is important as it is an opportunity to clearly introduce the group members and outline the key points to be delivered by the team. It is often best for one person to introduce the team and the order, rather than having each member of the team introduce themselves individually. As with any presentation, a clear and confident introduction to a group presentation can put you on the road to a successful outcome.

Each section may be allocated to a specific group member so that one member gives the Introduction and then passes on to the next member to deliver the next section, with one person being tasked with leading on the conclusion. This is the most common method of delivering a group presentation. The transition between presenters is the key to success, so it can be useful to practise this and to agree how to transition between speakers. Timing is vital – if the early speakers 'waffle' for too long, the last speaker will have no time left to deliver the conclusion on behalf of the group.

WHAT DO STUDENTS SAY?

'In groupwork, it always annoys me when some people don't pull their weight. We had quite a lot of group presentations on my course, so I started asking my group to take minutes of meetings so we had a record of who turned up and who wasn't contributing. We then used the records of these meetings in our assignments and attached them as appendices to our submitted work.'

(Ayoola, graduated with a BA (Hons) degree in Geography and Planning)

Handling questions and feedback

After a presentation, there are often questions and perhaps additional feedback from your tutor or others on your course. It can be useful to prepare for likely questions, as there may well be topics or issues that you know that people in the room will ask about – but do not be frightened to ask for clarification or further information if their question is unclear. It may be useful to ask for a specific part of a question to be repeated. Perhaps the person asking the question was not clear or did not speak very loudly. In such cases, it is better to ask for clarification than to try to guess what they are really asking. It can be useful to repeat the question yourself aloud, as this gives you time to think through the answer and is an opportunity for the person asking the question to confirm that you have understood them correctly.

Take time to think through your answer. A pause to think it through is more likely to come across as a calm, considered and professional response than blurting out a poorly thought-through response. In addition, if you do not know the answer, then it is best to admit it. Perhaps there is one part of the question that you can answer but another part that you are not able to, or are unsure how to respond. You cannot know everything, so admitting that you do not know can be useful in demonstrating that you are aware of the limits to your knowledge and understanding. It may also be that the question can be addressed by your tutor or seminar leader, or that you could find out the answer before the next class.

TOP TIP

If you are unclear about a question that you have been asked, ask for it to be repeated. You could also repeat the question yourself or reword it to ensure that you have understood what it is that you are being asked.

Summary points

1. Developing your presentation skills will not only help you with your Geography degree; these skills are crucial when it comes to securing employment after graduation.
2. In preparing a presentation, think through who your audience is, what materials you will need and its location.
3. Positively visualising your presentation as being a success can contribute to a positive outcome.
4. For a group presentation, discuss who will introduce it, who will lead on specific parts and who will answer questions. Consider the various strengths of the group members when allocating roles.

Key reading

Hill, J. & Walkington, H. (2016) 'Effective research communication.' In: Clifford, N., Cope, M., Gillespie, T. and French, S. (eds.) *Key Methods in Geography*. London: Sage.

CHAPTER 20

DOING HUMAN GEOGRAPHY FIELDWORK

Why undertake fieldwork?

The *Subject Benchmark Statement for Geography* tell us that Geography is fundamentally a field-based subject and that field experience is an essential part of geographical learning (QAA, 2019). Consequently, fieldwork is a key component of most Geography degrees and it is likely that you will carry out some in your first year at university. Some departments even offer an induction field trip that will see you conducting geographical research in the first couple of weeks of term. Your initial experiences of fieldwork at university are likely to be group-based and lecturer-led; these tend to be a part of a specific module or may count as a whole module, so the assessment related to fieldwork will form a part of the marks that you receive for your work. The groupwork involved in field trips will help to develop your communication and team-working skills and can involve presenting your findings to your peers (see Chapter 19 on presentations). Your lecturers and tutors will inform you of the details of any field trips that you will need to participate in during your Geography degree. However, as you proceed through your degree, there will be an expectation that you will work more independently, devising your own projects and conducting fieldwork as part of them (such as for your dissertation). As this suggests, an important transition over the course of your Geography degree will be a movement away from learning geographical knowledge to producing new geographical knowledge, too.

WHAT DO STUDENTS SAY?

'I was really nervous about going on my first field trip. I was living at home and so I hadn't met many people on my course compared to others who were staying in halls. However, I made so many new friends during that field trip – many of them remained friends throughout my degree and I am still in touch with them now that I have graduated.'

(Nazaneen, graduated with a BA (Hons) degree in Geography)

What we mean by geographical fieldwork can vary widely, ranging from the natural science approach more commonly found in Physical Geography (see Chapter 21) through to the social science and humanities approaches more commonly associated with Human Geography (which we will discuss further in this chapter). Although their approach may be very different, fieldwork is always a good thing to do – the Royal Geographical Society with the Institute of British Geographers has identified many personal benefits for us as geographers when we carry out geographical fieldwork, as noted in Table 20.1.

The academic purpose of doing fieldwork at university is ultimately about creating new knowledge related to the world around you – such as the exploration of little-understood events, to seek explanations of issues or to be able to describe, understand and predict what the future holds. Marshall and Rossman (1995, cited in Kitchin and Tate, 2000, p. 2) devised this handy table to help us to understand better the various academic purposes of doing geographical fieldwork.

So, we can see that fieldwork and the learning that takes place in the field enable us to develop our knowledge and understanding, as well as our skills, in relation to specific geographical issues and topics. More specifically, doing Human Geography fieldwork can help us to develop curiosity about and interest in the relationships between people and place and the role of society, the economy,

Table 20.1 Benefits of doing geographical fieldwork.

Broad educational purpose	Geographical fieldwork aim	Outcome for students
Conceptual	Developing knowledge and understanding of geographical processes and issues	• Improved academic achievement • A bridge to higher order learning • Opportunities for informal learning
Skills-related	Developing skills in data collection, presentation and analysis with real data	• Skills and independence in a wide range of contexts • The ability to deal with uncertainty
Aesthetic	Developing sensitivity and appreciation of built and natural environments	• Stimulation, inspiration and improved motivation • Nurture of creativity
Values-related	Developing empathy with views of others and care about/for the environment	• Developing as active and stewards of the environment
Social and personal development	Personal, learning and thinking skills such as independent enquiry, critical thinking, decision-making and teamwork skills	• Engaging and relevant learning • Challenge and the opportunity to take acceptable levels of risk • Improved attitudes to learning

Source: Marshall & Rossman (1995, cited in Kitchin & Tate, 2000, p. 2).

Table 20.2 Reasons for undertaking geographical fieldwork.

Exploration	To investigate little-understood phenomena. To identify/discover important variables. To generate questions for further research.
Explanation	To explain why forces created the phenomena in question. To identify why the phenomenon is shaped as it is.
Description	To document and characterise the phenomenon of interest.
Understanding	To comprehend and understand process, interaction, phenomenon and people.
Prediction	To predict future outcomes for the phenomenon. To forecast the events and behaviours resulting from the phenomenon.

Source: Marshall & Rossman (1995, cited in Kitchin & Tate 2000, p. 2).

culture, politics, the history of people and places and the importance of environmental change. In the next section, we focus in more detail on the process of doing Human Geography fieldwork.

What is 'the field' in Human Geography fieldwork?

The stereotypical image of fieldwork tends to be one of geographers standing in remote, muddy, rural places, carrying loads of equipment to survey the landscape while wearing yellow hi-vis jackets and boots. However, for Human Geographers, 'the field' can mean many things and be in a whole host of locations (not all remote).

For some Human Geographers, the field means off-campus (such as in wider society, such as in specific communities or cities); in the economy (such as in specific workplaces or institutions); in cultural venues (such as museums, galleries or theatres); or in places associated with politics and international relations (such as in parliament or local government offices). Consequently, an important part of Human Geography fieldwork is about negotiating access to these specific locations and establishing how willing specific groups or individuals are to be involved in your fieldwork. A key consideration must always be the resources available to you to complete your fieldwork effectively.

Doing Human Geography fieldwork does not always require us to work in specific places or with specific people. Often, it is possible to access our 'field' from home or from the university library. For example, for Historical Geographers, doing fieldwork can mean accessing document archives stored online by museums. For Population Geographers, doing fieldwork can involve studying secondary data, such as accessing census data or data from national attitudes surveys online.

TOP TIP

Always consider the resources needed, as well as the ethics and risks associated with doing Human Geography fieldwork.

Whichever field site you enter, there are important questions to ask yourself, not only about the research you will be doing but about your ethical conduct and your health and safety. For example, will you be travelling in a large group or in smaller teams? How will you negotiate access to specific places or people that you are interested in studying? How will you interact with the various people, places and communities in a way that is respectful to them and safe for you? Another important consideration upon completion of your fieldwork is how you leave your field and how you will minimise harm and promote high ethical standards in the process.

Doing Human Geography fieldwork: research aims, questions and hypotheses

Before undertaking any fieldwork, it is important: (a) to focus your project on a specific topic; and then (b) to identify research aims and questions associated with that topic which you will seek to address through your fieldwork. An important way of narrowing down your topic is to read relevant literature. Reading should help you to answer these important questions before you begin:

- What has and has not been researched on this topic already?
- Why am I doing this fieldwork?
- Why is this fieldwork needed?
- What is the gap in existing knowledge that my fieldwork will help to fill?

It can be difficult to come up with strong research questions, as these need to be about generating new knowledge and to be addressed by you, given the time and resources available. When formulating research questions, the general rule is: the more specific, the better.

TOP TIP

Every time you devise a research question, keep revising it and focusing it down so that you make it as specific as you can.

Some students think that all Human Geography fieldwork must be based on testing a specific hypothesis. A hypothesis is a proposition made about the topic that you are exploring, which you then seek to prove or disprove. This can be a useful way of approaching fieldwork, but it is only one way of doing so and it is by no means a requirement of Human Geography fieldwork to have a hypothesis. Many Human Geographers do not, and instead use research aims and/or research questions to seek to explore an issue or better understand a set of relationships.

Different types of data

One of the benefits of undertaking Human Geography fieldwork is that it enables you to appreciate various types of data and ways of collecting data to better understand the world around you.

Primary and secondary data

Primary data are new data that are collected at first hand in the field. In other words, these are the data that you collect when doing Human Geography fieldwork.

In contrast, secondary data are data that already exist. This could be, for example, census data, local government data, national health survey data or other such data as could help to inform your study. It can sometimes be useful to try to use both types in the same project – but other times this is not necessary. Often, a good starting point is to know what secondary data are available on your topic before you seek to collect primary data during fieldwork. There is little point in collecting data where there are already secondary data available that address the question or issue that you are studying.

Qualitative v quantitative data

Qualitative research gathers data that are non-numerical and often relate to people's options or perceptions of the world (such as diaries, interviews, focus groups and observations). On the other hand, quantitative research gathers data that can be analysed in a numerical form. It is often better at revealing patterns and trends than explaining why they occur. Examples of quantitative research include question-naires that used closed questions or rating scales to collect information. Note that questionnaires are often not the best way to collect qualitative data, as they offer little depth compared to interviews and focus groups.

WHAT DO STUDENTS SAY?

'My Geography teacher always used to tell us to show a range of skills by including some qualitative and some quantitative data, some primary and some secondary data, in our fieldwork projects. It took a while to get my head around the fact that at university the expected approach was to figure out the best type of data to answer the question and just collect that – even if, for example, it meant only doing qualitative interviews on a field trip.'

(Henry, graduated with a BA (Hons) degree in Geography)

Inductive and deductive

Inductive research is where the research comes first, and you then use it to generate a theory about the issue that you are researching. Deductive research is where there is already a theory and you seek to apply or utilise this during your fieldwork to explain what your results show.

When it comes to deciding on your approach to each of the above, the key is to select the approach that is most suitable for your study. It is impossible to do everything at the same time – the best fieldwork is that which adopts the approaches most suitable to the topic and issue being addressed. It may be that there is no secondary data on the topic that you want to study, or perhaps the secondary data that exist are quite old or not directly relevant. If this is the case, you should not worry about not using secondary data. Another important consideration when doing fieldwork is the methods that you will use. Again, this is not about using multiple methods, as some will be better than others in addressing the specific questions that you are asking.

Summary points

1. Undertaking fieldwork is a key part of your Geography degree and can serve many purposes, including developing conceptual and critical thinking skills and enhancing your social and personal development more generally.
2. Geographers undertake fieldwork for many reasons, including to explore, explain, describe, understand and predict geographical phenomena.
3. Before undertaking fieldwork, you should engage with the relevant literature and identify specific research questions that you will seek to address through your fieldwork.
4. The 'field' in Human Geography is diverse and means many things, depending on the type of research that you are planning to do.

Key reading

Clifford, N., Cope, M., Gillespie, T. & French, S. (2016) *Key Methods in Geography*. London: Sage.
Kitchin, R. (2000) *Conducting Research into Human Geography: Theory, Methodology and Practice*. Harlow: Prentice Hall.
Phillips, R. & Johns, J. (2012) *Fieldwork for Human Geography*. London: Sage.

CHAPTER 21

FIELD AND LAB RESEARCH IN PHYSICAL GEOGRAPHY

▶ By Dr Simon Drew

Doing Physical Geography fieldwork

The opportunity to do fieldwork is a big draw for many new Physical Geography students. Apart from the intellectual challenge of learning-by-doing, being outside exploring new places, meeting different people, being on glaciers, in rivers, in beautiful rural landscapes or (insert your favourite environment here) is the main reason why many students choose to study Geography. It can be uplifting and inspiring, and everybody has their favourite fieldwork stories. Although you will have done some Geography fieldwork at school or college, upon arriving at university you should find that you have considerably more opportunities to get out and experience fieldwork, travelling overseas or within the United Kingdom, taking peat cores, sampling water and sediments from rivers, mapping moraines and so on.

WHAT DO STUDENTS SAY?

'I loved fieldwork. For me, doing things made the degree more interesting than sitting in lectures and seminars all the time . . . . It was good fun, too.'

(Paige, graduated with a BSc (Hons) degree in Physical Geography)

While fieldwork is really good fun, one area that is taken very seriously by all universities is health and safety. *Please don't stop reading!* As universities have expanded and professionalised, this has become considerably more important, especially in Physical Geography, where you are more likely to end up in an environment or pursuing activities that may be quite dangerous (out on boats, climbing on cliffs, wading through rivers: the list is endless). When I was a postgraduate student trying to collect data for my thesis, I found myself in a sinking child's dinghy, half a mile offshore with no life jacket or phone, and no one knew where I was. Because of the systems that universities have in place, it is inconceivable that that would happen today. Most of the time these systems will be invisible to you, but it is important that you know that they are there.

TOP TIP

Don't buy any outdoor clothing, wellies, walking boots, and so on before you arrive at university. Some Geography departments have a supply of these that they lend out to students. If not, you'll probably get a discount at outdoor shops with your new Students' Union card.

Going on geographical expeditions

Universities are bursting with opportunities to have new experiences and to try new things. For many new Geography students, dipping their toe into this new world involves joining one or more university clubs and societies. After you have tried your students' union's Pot Soc (Pottery Society) or Cock Soc (Cocktail Society), you may want to consider looking at some of the opportunities to extend and develop your interest in geography.

So, what form do these opportunities take? Whether you see yourself as a Human, Physical or Environmental Geographer (see Chapter 2), probably one of the best opportunities available to you is to organise, and be part of, your own student-led expedition. Most universities have some sort of support for this, usually their Expedition Society. Some will make a contribution towards the cost of student-led expeditions, although many students fund their expeditions by applying for a grant from the Royal Geographical Society (RGS) or another philanthropic group with grants available. In my university, we have recently had groups go off on expeditions to explore things as diverse as: glaciers melting in Svalbard; mangrove biogeochemistry in Australia; the legacy for South Africa of hosting the 2010 football World Cup; nomadic tribespeople in Mongolia; and manatees in Venezuela. These expeditions often double up as data collection for the students' final-year dissertations. Quite apart from having a once-in-a-lifetime experience, students who manage to successfully organise and complete these expeditions develop an exceptional set of transferable skills, which makes them very employable graduates in either Geography or non-Geography-related jobs (see Chapter 4 for more guidance about the careers that Geography graduates go on to pursue). One or two even go on to become explorers themselves!

TOP TIP

Look on the Royal Geographical Society's website to see what expedition grants university students can apply for. The RGS also offers discounted membership to Geography students – this is ideal for those who want to support their studies and personal interest, or demonstrate their enthusiasm for Geography.

Lab work in Physical Geography

Most new students are unlikely to have been into a laboratory before to learn about geography, and it may be a pretty unfamiliar environment at first. Indeed, not all Physical Geography students end up studying in one – but at university, undertaking lab work can be a very important aspect of Physical Geography. One reason that lab work has grown in importance is the development of new lab-based techniques that can help answer longstanding and new geographical questions. Some of the techniques and machines can sound incredibly intimidating (high-pressure liquid chromatography, gas chromatography, mass spectrometry, etc.) and make use of ludicrously expensive technology. Don't be put off by this! You will never look more the part as a geographical scientist than when you are in a lab, decked out in a white lab coat, safety glasses and a pair of rubber gloves, as you wait for your core sediment to oxidise in hydrogen peroxide under a fume hood!

TOP TIP

When you start your degree, no one expects you to have any real experience of lab work. Remember that the equipment is provided for you to use – respect it and look after it, but don't let the cost put you off using it.

Some standard techniques are the mainstay of most university Geography departments. These include activities like sieving sediment and soil samples to get grain-size data or preparing sediment core samples to extract pollen and diatom remains (plankton), which can be used to reconstruct environmental histories. Most departments will have sieving equipment and, if you are lucky, automatic sieving machines to do a lot of the work for you (hand sieving can be physically hard, very noisy and, of course, very dirty). When it comes to analysing your cores, I would be amazed if there is a Geography laboratory in the country without a selection of light microscopes. You will use these to analyse slides of the pollen and other microfossils found in your cores. You can probably expect to find lower-power, stereoscopic microscopes, which can be used to identify aquatic invertebrates and plants. Another standard bit of lab kit is some sort of water-quality analysis equipment. At school, you may well have used a basic pH meter. You will have these at university, too, but they are more accurate and precise and (as you may expect) much more expensive.

The role of new mapping technologies

Geographers have always been interested in maps. They are central to what geographers do and new technological advances are having a profound impact on the kind of questions we can ask and answer. Mapping has been transformed by satellite systems, such as the Global Positioning System (GPS) and Eurosat. Simple GPS handsets that cost about £100 are accurate to within a metre. More sophisticated systems are accurate to less than a centimetre (but cost about £20,000). I recently asked one of the technical support reps from a company selling this gear why they could not compress it all into a phone. I was joking, but he replied with a straight face: 'We are working on it. There will be an announcement about this within the next year.' Of course, it is unlikely that your lecturers are going to let you loose, unsupervised, with a 20-grand bit of research kit in your first year. You will probably have to learn the basics of surveying with an old-fashioned level on a tripod!

Another area where technology is making a massive impact is the use of drones. Mapping using aerial photography from drones and using drones to carry scanning devices is improving fast. Drones are currently demonstrated to our first-year students, yet final-year students have regularly made use of them in their dissertations, unsupervised. The use of drones is expected to be a huge growth area in the economy over the next decade, so this is an exciting and interesting area to get into and to develop your skills.

WHAT DO STUDENTS SAY?

'Flying one of the department's drones to collect data for my dissertation was the highlight of my degree.'

(Ali, graduated with a BSc (Hons) degree in Geography)

New technologies for quantitative data analysis

It is generally well understood by Physical Geography lecturers that many new Geography students do not have a strong background in Mathematics or Statistics. Therefore, many new students find this aspect quite intimidating and do not feel very confident. If that is you, it is worth bearing in mind that no Physical Geography student will be able to get through a BSc degree without some sort of degree-level training in quantitative techniques, looking for patterns, differences and trends in space. These may include t-tests, f-tests, ANOVA (analysis of variance) and other more complex analysis. Of course, your lecturers will teach you how to use these techniques to analyse your data, but to be a successful Physical Geography student you have to be up for the challenge of learning these skills.

While learning how to use these techniques, you will be expected to learn to use the computer software programs in which they are applied. These may include standard Windows programs, like Excel and SPSS. However, in recent years there has also been a move towards using very powerful, open-source computer programs such as R and MATLAB, capable of carrying out all of these standard techniques

TOP TIP

Physical Geography at university is often taught in a scientific way. It can have a very different 'feel' to that of school or college, where it is often taught in a less scientific way, with a stronger focus on the impact of people on the environment (or the environment on people).

and much more. Using this software requires students to learn some basic programming, which creates a much more versatile and useful platform. Geography departments with the staff with these kinds of skills will routinely teach their students how to use these tools and techniques.

As this suggests, Physical Geography students are often dealing with what are now called 'big data'. In other words, they work with extremely large data sets that can be analysed to reveal patterns, trends and associations. This has been assisted by the introduction of remote sensing. This is the acquisition of information without making physical contact with the object that you are studying. Remote sensing is used in many areas of Physical Geography, including land surveying, hydrology, ecology, meteorology, oceanography, glaciology, geology and so on.

WHAT DO STUDENTS SAY?

'I hadn't really done any maths since I was 16 years old, so it was quite a struggle at first to get back into it. Some students had, and I think they found it less of a shock than I did! However, I liked Physical Geography, so I stuck at it and the maths side became easier . . . . I didn't like getting muddy and wet, so I even ended up using remote sensing data for my dissertation!'

(Kyle, graduated with a BSc (Hons) degree in Geography)

Summary points

1. The opportunity to do fieldwork in the United Kingdom and overseas is a big draw for many new Physical Geography students. Some organise their own university expeditions. This develops an exceptional set of transferable skills, which makes them very employable graduates.
2. Most new students are unlikely to have been into a laboratory to learn about geography – but at university, the lab work can be an important aspect of learning about Physical Geography.
3. Physical Geography students enjoy the challenge of learning how to use new technology and equipment, while also developing their maths and stats skills.

Key reading

Strahler, A. (2013) Introducing Physical Geography. Chichester: John Wiley & Sons.
Holden, J. (2012) *An Introduction to Physical Geography and the Environment*. Harlow: Pearson Education.

CHAPTER 22

MAKING THE MOST
OF FEEDBACK

What to expect

At the end of your Geography degree course, you will be asked to complete something called the National Student Survey (NSS). This is a list of questions about the various aspects of your university experience. It covers everything from the quality of the teaching and learning that you experienced, through to the enjoyability of your university's students' union and the value of the sports facilities. As students on every degree course are asked the same questions, one purpose of the NSS is to compare degree courses in British universities.

The good news is that many Geography departments score 90%+ in the NSS for the key question about students' 'overall satisfaction'. This means that, overall, the vast majority of Geography students in British universities are satisfied or very satisfied with their degree course. In contrast, the area in which most Geography departments have historically had to work hardest to respond to students' needs is in relation to the question about feedback. Generally, the message has been that students want more feedback on their work, more regularly. So, what is feedback and what forms can it take? While the answer to this question may seem obvious, as we saw in Chapter 8, there are many types of feedback available to Geography students at university, some more evident than others.

TOP TIP

To maximise your potential as a student, you need to recognise and act upon the various forms of feedback available to you throughout your degree.

Formal summative feedback

Summative feedback is the most easily recognisable form of feedback that you will receive throughout your Geography degree. In fact, it is so recognisable that many students assume that it is the only form that their feedback takes! Summative assessments and summative feedback on them 'sum up' the extent of your learning, and the mark that you are awarded contributes directly to your final degree classification (see Chapter 8). They include things like end-of-term or end-of-semester exams, projects and essays. Traditionally, the style of summative feedback has been formal – the handwritten comments on your written assignments, along with an official university feedback sheet completed by the marker. Today, online marking is increasingly being used. This takes the same form, except that the comments are typed by the marker or recorded as a short podcast instead of being handwritten.

Your summative feedback from the marker should provide enough words and phrases from the marking criteria to justify why you were awarded your mark, while also offering advice about how to improve. Sometimes this is known as feed-forward advice, as it takes the form of things that you should focus on improving in your next assignment. In the first year of Geography degrees, feed-forward advice often focuses upon general issues that apply to all Geography assignments, such as: whether you are reading sufficiently; whether the structure of your answer is logical; whether your answer is critical/evaluative enough; whether the spelling and grammar are good; and whether the referencing is correct.

In most university Geography departments, feedback on your assignments will come from someone in the same department. Often it will be the same person who set the assignment, though sometimes this is marked by a group of lecturers. Whichever is the case, this gives you a massive advantage when it comes to getting more feedback, as the person who marked your assignment will be easily accessible. In fact, something that most lecturers wish students would do more is talk to them about their feedback, as there is only so much that they can write in the margins of an essay or in the relatively small box on the official feedback form.

One of the challenges that you will face early in your degree is finding the courage to knock on your lecturer's office door to ask for more feedback or to clarify the feedback that they have given you. The earlier you can overcome that hurdle, the better – you will receive more feedback and your marks will improve as you figure out more quickly what is expected of you.

WHAT DO STUDENTS SAY?

'When I got a good mark in first year, I was so happy that I often didn't bother to read the feedback to find out why my assessment was good. If it was a bad mark, I often couldn't bring myself to read the feedback! It was only into second year when I began to think about how daft I'd been not to read and think about the feedback I'd received, to make sure I didn't make the same mistake again and so make my next assessments better.'

(Mashhood, graduated with a BA (Hons) degree in Geography)

Informal formative feedback

Occasionally, formative feedback is formal in its tone. For example, in the final year some departments provide formative feedback on drafts of students' dissertations. However, the style of formative feedback is usually informal and therefore harder to recognise. Nevertheless, it is just as important as summative feedback, as its goal is to help you to identify quickly your strengths and weaknesses and to improve the areas that need work. For example, have you ever asked a lecturer a question in a seminar or on a fieldtrip? Have you ever asked for advice on a draft of an essay? Or sent a lecturer an email about an assignment that you were about to hand in? Alternatively, have you ever been asked a question in class by a lecturer? Or has a demonstrator ever told you they liked the map you made in a GIS practical? If you have, then you have been the recipient of informal formative feedback, and it has hopefully improved your understanding of the material and consequently made your next assignment better than it would otherwise have been.

TOP TIP

Get into the habit of asking your lecturers for more feedback on your assignments. A quick chat with the marker can often make the strengths and weaknesses of your assignment clearer than written feedback. However, do not attempt to negotiate or haggle with your lecturer for a higher mark – this rarely changes anything and is never well received!

Another important type of formative feedback that is often overlooked is peer review. This is where you swap essays with a friend taking the same module and give each other feedback. This is useful because, while you can talk to lecturers at length about your assignment before you hand it in, often university Geography departments will not allow lecturers to read full drafts of an assignment before submission. An important aspect of university is developing independent thinking, so this is a line that many departments will not cross. You should definitely ask if this is the policy in your department in the first few weeks you are at university. If it is, the next best alternative is asking a course friend to read your essay, as we can often spot the strengths and weaknesses in others' work easier than in our own. We can then learn from their strengths and point out their weaknesses. What we cannot do is to copy parts of their assignment, as this would be plagiarism, which is a form of cheating (see Chapter 16).

There is a common theme running through many of the above examples of formative feedback – the need to ask for help. While summative feedback will be provided automatically by your department, many of the types of formative feedback available to students are given only when you ask for it. A key change from school or college to university is the need to take control of your own learning, and an important part of that is recognising when you need help and then being confident enough to ask for it. Our suspicion is that plenty of Geography students who still say in the NSS that they wanted more feedback throughout their degree either did not recognise formative feedback when they received it or were reluctant to ask for it. Please do not fall into either of these traps.

WHAT DO STUDENTS SAY?

'I had this great pastoral tutor in first year, who was originally from Northumberland. He always used to tell us "Shy bairns get nowt", which basically means that if you're too shy to ask for feedback from lecturers, you won't get it. There was tons of help available from lecturers, but it took me a while to realise that I was missing out by not asking for it. And actually, the one-to-one feedback I asked for was often the most helpful.'

(Mike, graduated with a BA (Hons) degree in Geography)

Summary points

1. Before you begin any assignment, find your Geography department's marking criteria or rubric. Use it both as a pre-submission checklist to make sure you are delivering what the marker wants and as a post-feedback means to interpret and understand the marker's feedback.
2. Always read your feedback, understand it and engage with it. It is natural to feel that feedback is harsh or unfair. However, if you want to do well in your Geography degree, you need to engage with it to figure out what makes a good assessment answer from your lecturers' point of view.
3. Every couple of months read all of your feedback in one go. Identify common or recurring weaknesses and plan how to improve them. This may include asking lecturers for advice or asking your tutor if the university has a study skills centre that you can visit.

Key reading

Weyers, J. & McMillan, K. (2011) *How to Succeed in Exams and Assessments*. Harlow: Pearson Education.

CHAPTER 23

KEEPING BALANCE AND MAINTAINING WELLBEING

Don't be too hard on yourself

We hope we have managed to convey throughout this book that doing a Geography degree is a really exciting thing to do. Geography broadens the mind: it helps us to understand why countries are the way they are; why the Earth's surface is shaped the way it is; and how the things we do affect others. The range of skills that you need to succeed as a Geographer is equally broad – everything from statistics and GIS, to critical thinking and interviewing the public. Apart from being enjoyable, this contributes to making Geography graduates incredibly employable in professions as diverse as flood management, energy, teaching, accountancy, advertising and real estate.

We hope we have managed to convey that some aspects of your Geography degree will build on the knowledge and skills you acquired at school or college. Other aspects will be completely new to you. While exciting in many ways, this can understandably be a cause of concern. Will I enjoy it? Am I clever enough to do it? Will I be able to cope with the new demands? These are questions that many new Geography students ask themselves in their first few weeks at university.

TOP TIP

Beginning university involves both an academic transition and a social transition. This is a lot to adjust to, so give yourself a chance to settle in.

These doubts are not helped by the fact that degrees start at a time of personal change – leaving behind old friends and making new ones, moving into a hall of residence, starting a relationship, becoming more independent of parents and guardians, and managing time and finances in a way that you have not done before. This is a lot of change for anyone to get their head around, and it is important to see the transition to university as a process, not an event. In other words, do not expect to feel fully 'at home' by the end of induction week. This is unrealistic. Your lecturers will not expect it and you should not, either. Give yourself time to settle in. As we said in Chapter 1, we believe that beginning university involves both an academic transition and a social transition – and that there is a strong relationship between the two. In other words, if one aspect is not going well, you cannot be sure that the other aspect will go well either. So, in the first few weeks, prioritise making friends and settling in, as well as getting to grips with your new course.

Stress and anxiety

Stress means different things to different people, and it can involve a wide range of emotions and feelings, though often it involves some form of pressure that results in worrying or fretting. Most people feel stressed at some point in their lives and often find ways to cope. For example, no one likes exams, but we know they will be over in an hour or two and so generally we find a way through them – even if the stress means we do not always perform to the best of our ability.

University can also generate new situations that can make students feel stressed for longer periods of time, for example: too many things being demanded of you at the same time, such as multiple essay deadlines combined with uncompromising shift patterns at work; nagging doubts that you will not achieve your own expectations; feeling overwhelmed by the volume of reading expected each week; or feeling homesick. This type of stress is much harder to deal with and, unlike exam-related stress, most new students do not have tried-and-tested coping strategies that they can draw upon. This can lead to anxiety, which is one of the many adverse effects of stress. Anxiety is the process during which a person becomes

Table 23.1 The stress calendar.

First few weeks of the academic year	Feeling homesick Difficulties in fitting in Developing new relationships Encountering new teaching styles Questioning if you picked the right subject to study Feelings of academic inferiority
Mid-term or mid-semester	First assignment due to be handed in Lack of 'real' friendships Financial problems Difficulty in balancing social life, studies and the need to work Self-doubt
End of term or end of semester	Time-consuming extra-curricular activities (e.g. football team, volunteering) Lack of sleep due to end-of-module assessments, unsocial employment hours and end-of-year parties End-of-module exam revision Worsening financial problems Blues of coming home (if you are enjoying the freedom of living away from home)

Source: McMillian & Weyers (2009).

scared and apprehensive of what lies ahead, and often manifests itself in physical problems like pain, dizziness and panic attacks.

Factors in your personal life that may seem distant from your academic work can also be a source of stress. These might be connected to family issues, such as the death of a grandparent, one of your parents moving to a new place or changing jobs or a sibling or other relative being unwell or struggling with life. Closer to home, since your relationships with friends and family often change when you move to university (this can change even when you do not move away), it can mean that close friendships and relationships come to an end as you transition into life as a university student (for example, you might break up with a long-term romantic partner). On top of this, there are financial issues as you may find that university involves financial pressures that can be an additional burden on you. Added to this, you may have specific disability, health or medical concerns to consider, and there will be people at university who can advise you about how to access support.

Your personal values and identity play an important role here. Perhaps you identify with a specific youth subculture, have excelled at sport or are affiliated with a specific religious community or group. Some students find that moving to university enables them to make friends with new people who have similar interests, but this may take time as you find out about groups on campus or near to where you are studying. Some students find that moving to university gives them a chance to 'come out' about their sexuality or to explore new forms of gender expression. As you will see here, there is often a lot going on when you go to university and this is not only about lectures, seminars and assignments.

Responding positively to stress

As a new Geography student, one of the things to be wary of is becoming trapped in a downward spiral of stress. It often looks something like this: you feel under pressure to do well in your Geography degree; you study for long hours in the library; you feel tired, stressed and as though you do not have enough time to study and make many good friends; you do not do as well as you (unrealistically) expected in your first assignment; you study for even longer hours in the library; you feel even more tired and stressed … and so the cycle continues. Looking from the outside, it is easy to see why this spiral happens. It is easy to see that it is unsustainable and, if faced with this situation, you need to do something to manage your stress.

Luckily, if do you start to feel stressed, there are lots of positive things that you can do to try to manage it. Here are just a few.

Take regular breaks
Long periods of time spent studying do not necessarily mean better marks – particularly if you are not using this time efficiently. To achieve more in less time, use the tips in this book about studying effectively.

Share your concerns
Talking to friends or your tutor (see Chapter 7) can be a great source of support. Realising that others feel the same can often help to reduce stress levels.

Learn to prioritise
If you have loads of things to do, putting together a list can help you to break tasks down into manageable chunks, to prioritise tasks and to allocate time to each.

Avoid perfectionism
You need to do your best at university, but no assignment that you submit at university will be perfect. Accepting this can often help to reduce your stress levels.

Confront your problems
Avoiding a problem only reduces your stress temporarily. For example, if you leave assignments until the last minute you need to try to understand why you do this, and start work on them earlier.

Do some physical activity

This does not necessarily mean running a half-marathon or lifting weights in a gym. It could be as simple as going for a short walk or taking a yoga class. Activity releases endorphins and is a great way to reduce stress.

Sleep and eat well

To work at its best, your body needs time to refuel and recover. Having 8 or 9 hours of sleep each night, coupled with a balanced diet, can be a great way to manage stress.

Balance your studies, employment and social life

If any of these becomes overly dominant, you will not perform to the best of your ability at university. Getting the balance right is the key to long-term stress management, happiness and a sustainable lifestyle. For example, talk to your university about its hardship fund if you are working at a job for too many hours to study effectively.

WHAT DO STUDENTS SAY?

'I had no idea how to do well at university, so in the first few weeks I spent hours and hours in the university library studying and trying to read everything that lecturers mentioned. It just wasn't sustainable, I felt rubbish by Christmas and didn't do that well in my January exams. It wasn't until Easter that I felt I'd managed to get a better balance by joining the Geography netball team. I was spending less time studying but, by studying more effectively, my marks actually improved!'

(Pippa, graduated with a BA (Hons) degree in Geography)

Asking for help

With a few relatively minor hiccups along the way, most new Geography students make the transition to university successfully and go on to complete their degree. However, some will need to ask for help. This could be because: the tips above have not reduced their stress levels; because their stress has turned to anxiety, depression or another mental health condition; because of enduring homesickness (see also Chapter 1); or because they feel overwhelmed by the thought of doing a degree.

TOP TIP

Don't feel ashamed about admitting that you are struggling to cope, and instead ask for the help you need.

Universities recognise that students respond differently to the academic and social challenges associated with settling into university, and have a range of support available – please use this, as the earlier you seek support by talking to people about how you are feeling, the easier it will be for them to offer you effective help. For example:

Contact your tutor

An important part of your tutor's role is to listen to any concerns you have and to offer support and advice (see Chapter 7). Also, ask your tutor for an extension to an assessment deadline if you feel that a short break from your studies will help you to manage your stress levels.

Contact your university's counselling service

Your tutor may advise you to contact this service, which is staffed by professionals. They will make you feel at ease, help you to work through your problems and put you in contact with others who can help you. This service is confidential and separate from your Geography department. Your lecturers will never find out that you have been to the university's counselling service, let alone what you said.

Summary points

1. Have realistic expectations. Beginning university involves both an academic transition and a social transition. This is a lot to adjust to, so give yourself a chance to settle in.
2. Looks for ways to positively manage your stress levels on a day-to-day basis.
3. Universities are not interested in supporting you only academically throughout your degree. Never be afraid to ask for help with your wellbeing when you need it.

GLOSSARY OF KEY TERMS

Aims Each degree programme will have a set of aims that explain the overall goals of the programme. These aims will relate to programme structure, student outcomes, placements (where relevant) and accrediting bodies (where relevant). Modules will also have a set of aims that explain the primary objectives of each specific module.

Assessment A generic term for a set of processes that measure a student's achievement of the intended learning outcomes in terms of knowledge acquired, understanding developed, skills gained and attributes demonstrated.

Assessment criteria Descriptions by which an assessor determines whether a student has demonstrated the achievement of the intended learning outcomes for a particular level.

Assessment methods The different means by which a student's achievement of the intended learning outcomes can be assessed. A wide range of methods may be used, but they must be appropriate to the intended learning outcomes being assessed.

Bachelor's degree A degree usually obtained after three years or more of full-time study. Human geographers can graduate with a Bachelor of Arts degree (BA) or a Bachelor of Science degree (BSc). Physical geographers normally only graduate with a Bachelor of Science degree (see also undergraduate and postgraduate).

Campus A campus is the land on which a university is situated. Usually it includes libraries, lecture halls, halls of residence, a students' union building, departmental buildings and admin buildings.

Credit A quantitative measure of learning effort. The size of a module is measured by reference to student learning time, so that for every 10 credits a student is expected to spend 100 hours in programmed activities, private study or assessment. Credit is normally awarded for the achievement of a set of specified intended learning outcomes.

Degree classification A means of distinguishing between the differences in achievement by individual students of the intended learning outcomes for a degree programme. Typically, the degree classifications awarded include: a first-class degree (1st); an upper-second class degree (2:1); a lower second-class degree (2:2); a third-class degree (3rd).

Ethnography A qualitative approach to research that involves in-depth exploration of specific customs, cultures or communities.

External examiner An academic from another institution who checks and validates the marks and grading of degree awards, in consultation with the institution's own examination board.

Faculty A faculty group of university departments concerned with a major division of knowledge. Universities group their departments in different ways, but examples of faculty names include the Faculty of Humanities and Social Science, Faculty of Life Sciences and Faculty of Arts.

Feedback The process by which students are informed of their strengths and weaknesses. The aim should be for high-quality and timely feedback to enable students to assess their progress and to improve upon it.

Freshers First-year students in their first few weeks at university.

GIS Geographic information system. GIS is software that maps specific quantitative data patterns.

Graduand and Graduate A graduand is a student who has completed their course but who is still waiting to receive their degree. For most students, this period only lasts a few weeks. A graduate has successfully completed a degree course and has already been awarded their degree.

Halls of residence Most universities provide halls of residence. They are usually furnished flats, with several single bedrooms sharing a kitchen, toilet, bathroom and lounge area. Some are catered, others part-catered or self-catered. Some are mixed sex, others are single sex.

Hardship fund Fund administered by a UK university or college, making small payments to students with financial difficulties.

Honours Bachelor's degrees are generally awarded as honours degrees in one of three classes, First, Second or Third, depending on the overall marks awarded and the number of credits gained.

Joint honours A type of degree where a student studies two subjects in equal depth, for example Geography and Planning.

Learning outcomes Each programme will have a set of learning outcomes that specifies the skills and knowledge that students are expected to develop over the course of the programme. Modules will also have specific skills outcomes and knowledge outcomes that specify what you will learn and what skills you will develop on each module.

Lecture A lecture is a presentation intended to convey information about a topic. Traditionally, the lecturer will stand at the front of the room and recite information using some form of visual aid such as PowerPoint. Lectures can also involve a range of media such as videos and student participation in different formats. Expect your Geography lectures to range in size from a handful of students to a few hundred.

Master's degree A master's degree is the first level of postgraduate study. To apply for a master's degree, students usually must already hold an undergraduate degree (a bachelor's degree). A master's degree typically requires a year or two years of full-time study. It is usually abbreviated to MA or MSc, depending on whether the student focused on an arts/humanities or a science-related subject.

Matriculation This is the formal process for registering or enrolling at university. You often need to matriculate every year. Some universities refer to this as registration or enrolment.

Module An element within a programme of study. Many modules focus on one aspect of Geography in detail, such as *Geomorphology*, *Social Geography*, or the *Dissertation* module. How much each module contributes to your degree varies and is normally stated in credits/points. Many UK undergraduate academic years consist of 120 credits and your total study time is expected to be 100 hours for each 10-credit module. It can be useful to check to see how many credits each of your modules is worth and how many credits you need to pass each year.

- Compulsory modules – modules that you must take in order to fulfil the requirements of the degree programme.
- Optional modules/electives – modules which you choose to take because they suit your interests and career aspirations.

Module leader The member of academic staff responsible for the module is the module leader. If you encounter difficulty in a module, you should refer to the module leader in the first instance. If you cannot resolve the difficulty with the module leader, you should consult the degree programme director/degree scheme leader.

NUS The National Union of Students. If an institution's students' union is affiliated to the NUS, its students are automatically members of the NUS.

Pass degree Students who are unsuccessful at the end of their honours course may be permitted to take reassessments to attempt to achieve a pass at honours level. They will then be awarded a pass degree.

PhD Doctor of Philosophy. A PhD is a degree awarded to people who have completed advanced original research, over a period of three or more years, into a particular subject. Many of the academic staff at university will have a PhD in Geography or in a related subject area. In some universities, postgraduate students who are studying for a PhD will be involved in teaching undergraduate students, especially through seminars and tutorials.

Plagiarism Plagiarism is passing off someone else's work, whether intentionally or unintentionally, as your own for your own benefit.

Postgraduate student A student who is continuing their university studies 'post' (i.e. after) their first graduation. In other words, a student who has completed their undergraduate degree and is now studying for a higher degree, such as a master's degree or PhD.

Poststructuralism An approach to understanding the world that rejects specific 'truths' or 'facts' in favour of different opinions and different perspectives.

Prerequisite A module that is required to be studied before undertaking a further module that assumes prior knowledge.

Programme specification A comprehensive description of all features of a programme of study, including the intended learning outcomes, the means by which those outcomes are achieved and demonstrated, the curriculum, criteria for admission, student support and regulations for assessment.

Qualitative research Qualitative research tends to focus on people's values, behaviours and feelings and so involves the analysis of data about people's opinions or lived experiences.

Quantitative research Quantitative research employs data that can be numerically quantified and often involves statistical techniques or computational work.

Reading list List of the books that students are expected to read for a particular course.

Reading week A week during the academic year when students are expected to concentrate on reading and studying for their course. There are usually no lectures or seminars during this period. Not all universities have reading weeks; some courses may have reading weeks, others do not.

Semester A semester is a block of time in the undergraduate teaching year, typically lasting for between 12 and 18 weeks. Some modules will run in only one semester; other modules will run in both semesters. At many universities, both semesters end with a period of examinations. Many universities have both semesters and terms (and sometimes these names are used interchangeably).

Seminar A small class where students discuss a specific topic with a lecturer or tutor. Seminars can take several formats (see Chapter 6); some may be led by the lecturer or tutor and, on other occasions, students may be expected to lead the seminar.

Subject benchmark statements A description of the nature and characteristics of programmes of study in a specific subject. Subject benchmark statements represent general expectations about the standards for the award of qualifications at a given level, and they articulate the attributes and capabilities that those possessing such qualifications should be able to demonstrate.

Term The academic year at most UK universities runs from September or October to June or July, with a three- or four-week break around Christmas and Easter that divides the year into three terms. Some older universities have special names for the three terms, e.g. Michaelmas. Many universities have both terms and semesters.

Transcript A summary record of a student's academic achievements on a particular programme of study.

Tutor An academic staff member or teacher who is the main point of contact for a small group of students (tutees). A student may have several types of tutor (for example, an academic tutor and a pastoral tutor). Tutors are normally the first point of contact for any general queries a student may have about studying Geography at university.

Tutorial In some universities, the terms 'seminar' and 'tutorial' are used interchangeably. In other universities, tutorials involve only a very small group of students meeting with a tutor to discuss an academic topic. Alternatively, tutorials can refer to individual pastoral meetings between a tutor and a student to talk about their progress and wellbeing.

Undergraduate A student who is studying for a bachelor's degree.

REFERENCES

Cottrell, S. (2017) *Critical Thinking Skills: Effective Analysis, Argument and Reflection*. London: Palgrave Macmillan.

Day, T. (2013) *Success in Academic Writing*. London: Palgrave Macmillan.

Kitchin, R. &Tate, N. (2000) *Conducting Research in Human Geography: Theory, Methodology and Practice*. Abingdon: Routledge.

Maier, P., Barney, A. &Price, G. (2009) *Study Skills for Science, Engineering and Technology Students*. Harlow: Pearson Education.

McMillan, K. & Weyers, J. (2009) *The Smarter Study Skills Companion*. London: Pearson.

McMillan, K &Weyers, J. (2012) *The Study Skills Book*. London: Pearson.

Neville, C. (2007) *The Complete Guide to Referencing and Avoiding Plagiarism*. Maidenhead: Open University Press.

Oxford English Dictionary (OED) (1993) *The New Shorter Oxford English Dictionary*. Oxford: Oxford University Press.

Open University (2013) *Skills for OU Study: Thinking Critically*. Milton Keynes: Open University.

Oxford Brookes University (2018) *Plagiarism – What is It?* Available at https://www.brookes.ac.uk/library/library-services/information-skills/plagiarism (accessed 13 August 2019).

Quality Assurance Agency (QAA) (2019) *Subject Benchmark Statement for Geography*. Gloucester: Quality Assurance Agency for Higher Education.

Royal Geographical Society (RGS) (2019a) *What is Geography?* Available at https://www.rgs.org/Geography/what-is-Geography (accessed 8 August 2019).

Royal Geographical Society (RGS) (2019b) *Fieldwork Strategies*. https://www.rgs.org/CMSPages/GetFile.aspx?nodeguid=5e533b29-ad18-4de4-8775-d5b1e2a18868&lang=en-GB (accessed 15 September 2019).

University College London (UCL) (2019) *Exam Success Guide*. https://www.ucl.ac.uk/students/exams-and-assessments/exams/exam-success-guide#10%20steps (accessed 27 August 2019).

University of Leicester (2019) *Essay Terms Explained*. https://www2.le.ac.uk/offices/ld/resources/writing/writing-resources/essay-terms (accessed 27 August 2019).

INDEX

Note: Information in figures and tables is indicated by page numbers in *italics* and **bold**, respectively.

Printed in the United States
By Bookmasters